Engineering Interrelated Electricity Markets

Anke Weidlich

Engineering Interrelated Electricity Markets

An Agent-Based Computational Approach

Physica-Verlag
A Springer Company

Dr. Anke Weidlich
University of Mannheim
Dieter Schwarz Chair
of Business Administration
and Information Systems
Schloss
68131 Mannheim
Germany
weidlich@uni-mannheim.de

ISBN: 978-3-7908-2067-6   e-ISBN: 978-3-7908-2068-3

DOI: 10.1007/978-3-7908-2068-3

Library of Congress Control Number: 2008934063

© 2008 Physica-Verlag Heidelberg

This work is subject to copyright. All rights are reserved, whether the whole or part of the material is concerned, specifically the rights of translation, reprinting, reuse of illustrations, recitation, broadcasting, reproduction on microfilm or in any other way, and storage in data banks. Duplication of this publication or parts thereof is permitted only under the provisions of the German Copyright Law of September 9, 1965, in its current version, and permission for use must always be obtained from Springer. Violations are liable to prosecution under the German Copyright Law.

The use of general descriptive names, registered names, trademarks, etc. in this publication does not imply, even in the absence of a specific statement, that such names are exempt from the relevant protective laws and regulations and therefore free for general use.

Cover design: WMXDesign GmbH, Heidelberg

Printed on acid-free paper

9 8 7 6 5 4 3 2 1

springer.com

Preface

This book is a result of my work as a research assistant in the Information & Market Engineering group within the Institute of Information Systems and Management at the University of Karlsruhe, and later at the Dieter Schwarz Chair of Business Administration and Information Systems, University of Mannheim. Many ideas contained in it were generated within the PowerACE project, which was carried out in cooperation with the Institute for Industrial Production and the Fraunhofer Institute for Systems and Innovation Research, both in Karlsruhe.

Many people made the completion of my dissertation possible and should be acknowledged here. First, I am greatly indebted to Prof. Dr. Christof Weinhardt who gave me the opportunity to take part in his dynamic research team and supported me all the way through my dissertation. Next, I thank Prof. Dr. Daniel Veit for his continuing encouragement, guidance and strong support during our joint work. He gave me tremendous opportunities; interacting with him at his new chair was both instructive and pleasurable. For her inspiration as well as for her great hospitality, my appreciation also goes to Prof. Dr. Leigh Tesfatsion from the Economics Department at Iowa State University, who enriched my understanding of Agent-Based Computational Economics and whose great dedication to science impressed me strongly.

A final thanks goes to my colleagues in both Karlsruhe and Mannheim, who created the constructive and comfortable atmosphere that I so enjoyed, and whose critique and assistance contributed much to my work. Among my colleagues and friends, I would especially like to thank Dr. Clemens van Dinther, Carsten Block and Arne Weiß for proofreading the manuscript and, along with Jun.-Prof. Dr. Stefan Seifert, for giving me valuable advice during the finalization of my dissertation.

Karlsruhe/Mannheim, April 2008 *Anke Weidlich*

Foreword

After liberalization, power markets have been subject to rapid changes: electricity is traded at exchanges, new intermediaries come into place, and the regulatory authorities make strong efforts to protect consumers and, at the same time, meet the requirements of climate protection. The introduction of CO_2 emissions trading is an essential step in approaching the latter challenge. The resulting allowance markets, though, have an effect on electricity markets and vice versa. The complexity of these markets, and their interrelations in particular, make planning a difficult problem for all trading participants.

Large utilities, smaller power producers, and consumers must adapt to the new framework set by these developments. Simultaneously, legislators must learn to better understand how certain interventions affect strategies and prices in the market, both from the perspective of the supply side and the demand side. In this context, there is a growing need for adequate models and tools that help decision makers to decide on appropriate measures for coping with their challenges. Complex systems call for flexible modeling approaches, such as the agent-based methodology applied in this book. Agent-based simulation is a promising methodology for supporting science-based engineering of markets in the electricity sector.

This book presents a simulation model which realistically represents daily wholesale electricity trading and emissions trading, and which can be applied to derive conclusions on how to best engineer these markets. In particular, the interrelations that exist between the power exchanges and daily markets for balancing power on the one hand, and emissions trading on the other, are smartly modeled by applying reinforcement learning and considering the generator's opportunity costs. The research results presented in this book provide an essential contribution to theory and practice. The author answers important prevailing questions, and also critically scrutinizes the employed methods as well as the resulting simulation outcomes and their implications on electricity markets.

We wish this book success, and insightful pleasure to its many readers.

Universität Karlsruhe (TH), May 2008 *Prof. Dr. Christof Weinhardt*
Universität Mannheim *Prof. Dr. Daniel Veit*

Contents

List of Abbreviations xiii

List of Figures xv

List of Tables xxi

Part I Motivation and Fundamentals

1 **Introduction** 3
   1.1 Objectives and Research Questions 4
   1.2 Thesis Structure 5

2 **Wholesale Electricity and Emissions Trading** 7
   2.1 Electricity Marketplaces 8
      2.1.1 Power Exchanges 8
      2.1.2 Balancing Power Markets 11
   2.2 Emissions Trading 14
   2.3 Interrelations Between Day-Ahead, Balancing and Allowance Markets 17
   2.4 Summary 19

3 **Agent-Based Computational Economics** 21
   3.1 Motivation for Agent-Based Methods in Economics 22
   3.2 Building Agent-Based Simulation Models 25
      3.2.1 Methodology and Main Concepts 25
      3.2.2 Validity of Agent-Based Simulation Models 29
      3.2.3 Software Toolkits for Agent-Based Simulation 31
   3.3 Related Work: ACE Electricity Market Models 33
      3.3.1 Simulations Applying Reinforcement Learning 33
      3.3.2 Simulations Applying Evolutionary Concepts 39
      3.3.3 Other Agent-Based Electricity Market Simulations .. 42
      3.3.4 Discussion of ACE Electricity Approaches 49
   3.4 Summary 55

Part II An Agent-Based Simulation Model for Interrelated Electricity Markets

4 Representation of Learning and Adaptation 59
    4.1 Reinforcement and Belief-Based Learning 60
        4.1.1 Erev and Roth Reinforcement Learning 62
        4.1.2 Q-Learning 64
        4.1.3 Experience-Weighted Attraction 64
    4.2 Evolutionary Learning Models 65
    4.3 Analysis of Learning Algorithms for Agent-Based Simulations ... 67
        4.3.1 Criteria for Choosing a Learning Model 67
        4.3.2 Simulated Scenario 70
        4.3.3 Results 72
        4.3.4 Implications for Robust and Valid Agent-Based Simulations 81
    4.4 Summary 83

5 The Electricity Sector Simulation Model 85
    5.1 Design of the Simulation Model 85
        5.1.1 Overall Model Structure 86
        5.1.2 The Day-Ahead Electricity Market Model 89
        5.1.3 The Balancing Power Market Model 92
        5.1.4 The Model of Emissions Trading 94
        5.1.5 Learning Reinforcements and Market Interrelations 96
        5.1.6 The Graphical User Interface 97
    5.2 Validation 99
        5.2.1 Simulation Results of the Reference Scenario 100
        5.2.2 Sensitivity Analysis 106
    5.3 Summary 116

Part III Application, Evaluation and Discussion

6 Simulation Scenarios and Results 121
    6.1 Impact of Tendered Balancing Capacity on Electricity Prices 121
        6.1.1 Motivation 121
        6.1.2 Results 123
    6.2 Settlement Rules on the Balancing Power Market 126
        6.2.1 Motivation 126
        6.2.2 Results 127
    6.3 Increasing Supply Side Competition Through Divestiture 130
        6.3.1 Motivation 130
        6.3.2 Results 131
    6.4 Policy Implications and Summary 135

7 Conclusion and Outlook ... 137
7.1 Original Contribution ... 137
7.2 Outlook on Possible Future Work ... 140

Appendix

A Learning Model Testing Scenarios ... 141
A.1 Definition of the Two-Dimensional Action Domain and Spillover of Reinforcement ... 141
A.2 Simulation Results for Appropriate Learning Variants ... 143

B Reference Scenario ... 147
B.1 Demand Side Data Input for the Day-Ahead Market ... 147
B.2 Supply Side Data Input: Generators and Plants ... 148
B.3 Agent Characteristics on the CO_2 Market ... 149
B.4 Simulation Results for the Reference Scenario ... 150
B.5 Confidence Intervals of Sensitivity Analysis ... 159

C Statistical Analysis of Market Scenarios ... 161
C.1 Tendered Balancing Capacity ... 161
C.2 BPM Settlement Rule ... 162
C.3 Divestiture Scenarios ... 163

References ... 167

List of Abbreviations

AB	Agent-based
ACE	Agent-Based Computational Economics
BPM	Balancing power market
CO2M	CO_2 emission allowance market
CV	Coefficient of variation
DAI	Distributed artificial intelligence
DAM	Day-ahead market
ECX	European Climate Exchange
EEX	European Energy Exchange
EnWG	Energiewirtschaftsgesetz (German energy industry act)
EUA	European Emission Allowance
EU-ETS	The European Union's emissions trading scheme
EWA	Experience-weighted attraction
GA	Genetic algorithm
GUI	Graphical user interface
ISO	Independent system operator
LCS	Learning classifier system
LSE	Load serving entity
MAS	Multi-agent systems
MR	Minute reserve
MRE	Modified Roth and Erev reinforcement learning algorithm
NAP	National allocation plan
OTC	Over the counter
RL	Reinforcement learning
RNS	Random number seed
TSO	Transmission system operator
UCTE	Union for the co-ordination of transmission of electricity
VDN	Verband der Netzbetreiber e.V. (German union of electricity transportation system operators)

List of Figures

1.1	Structure of the dissertation	5
2.1	Frequency control as instructed by UCTE (2006)	12
2.2	Settlement prices for CO_2 futures and spot contracts of the first EU-ETS period (2005–2007)	16
2.3	Settlement prices for CO_2 futures contracts of the second EU-ETS period (2008–2012)	16
2.4	Day-ahead electricity and emission allowance spot settlement prices	18
3.1	Methodological process of agent-based simulation (Drogoul et al., 2003)	27
3.2	Procedure for studying the output of agent-based simulations (Windrum et al., 2007)	30
3.3	Transmission system in the model described by Krause et al. (2005)	38
4.1	Agent–environment interaction in reinforcement learning (Sutton & Barto, 1998)	61
4.2	The agent's state domain (with state indices)	71
4.3	Sequence diagram of the daily trading process on the day-ahead market	72
4.4	Average course of prices with different learning algorithms (simplified scenario, averaged over 50 simulation runs)	75
4.5	Average course of prices with Erev and Roth learning for different action domains	77
4.6	Average agents' bid prices as a function of marginal generation costs	78
4.7	Average course of prices with Experience-Weighted Attraction (simplified scenario)	80

5.1	UML class diagram of agents in the simulation model	86
5.2	Flowchart of the actions performed at each simulation iteration	87
5.3	Illustrative supply and demand curves at the day-ahead market	91
5.4	Graphical user interface of the simulation model: generator settings panel	98
5.5	Graphical user interface of the simulation model: further settings panels	99
5.6	Merit order of installed generating capacity for the reference scenario	101
5.7	UCTE load for 2006	101
5.8	Simulated and empirically observed prices at the day-ahead and balancing power market, January 2006, with Q-learning	102
5.9	Simulated and empirically observed prices at the day-ahead and balancing power market, September 2006, with Q-learning	103
5.10	Prices for all observations at the day-ahead and balancing power market, sorted by corresponding system load (Q-learning simulations)	108
5.11	Prices for all observations at the day-ahead and balancing power market, sorted by corresponding system load (original Erev and Roth RL simulations)	108
5.12	Prices for all observations at the day-ahead and balancing power market, sorted by corresponding system load (modified Erev and Roth RL simulations)	109
5.13	90% confidence intervals for $E(Z_j)$, comparison of Q-learning and original Erev and Roth RL	110
5.14	90% confidence intervals for $E(Z_j)$, comparison of Q-learning and modified Erev and Roth RL	110
5.15	Impact of the range of possible DAM actions on day-ahead electricity prices (Q-learning simulations)	111
5.16	Impact of the range of possible DAM actions on minute reserve prices (Q-learning simulations)	112
5.17	Relative differences of simulated prices with varying DAM action domains	112
5.18	Relative differences of simulated prices with varying portfolio integration parameter	114
5.19	Impact of CO_2 emissions trading on day-ahead electricity prices (Q-learning simulations)	115
5.20	Impact of CO_2 emissions trading on minute reserve capacity prices (Q-learning simulations)	116

6.1	Monthly/yearly average prices on the day-ahead (*left*) and balancing power market (*right*) for different tendered minute reserve quantities..	123
6.2	Impact of tendered minute reserve quantities on day-ahead market prices (Q-learning simulations)	124
6.3	Impact of tendered minute reserve quantities on balancing power market prices (Q-learning simulations).........................	125
6.4	Impact of BPM settlement rule on minute reserve capacity prices (Q-learning simulations).....................................	128
6.5	Impact of BPM settlement rule on day-ahead electricity prices (Q-learning simulations).....................................	129
6.6	Monthly/yearly average prices on the day-ahead (*left*) and balancing power market (*right*) for different divestiture scenarios	132
6.7	Impact of plant divestiture on day-ahead electricity prices (Q-learning simulations).....................................	133
6.8	Impact of plant divestiture on minute reserve capacity prices (Q-learning simulations).....................................	134
A.1	Action domain at the day-ahead electricity market with action indices	142
B.1	Simulated and empirically observed prices at the day-ahead and balancing power market, January–April 2006, with Q-learning, ε-greedy action selection, learning rate $\alpha = 0.5$, discount rate $\gamma = 0.9$, $\varepsilon = 0.2$..	150
B.2	Simulated and empirically observed prices at the day-ahead and balancing power market, May–August 2006, with Q-learning, ε-greedy action selection, learning rate $\alpha = 0.5$, discount rate $\gamma = 0.9$, $\varepsilon = 0.2$..	151
B.3	Simulated and empirically observed prices at the day-ahead and balancing power market, September–December 2006, with Q-learning, ε-greedy action selection, learning rate $\alpha = 0.5$, discount rate $\gamma = 0.9$, $\varepsilon = 0.2$	152
B.4	Simulated and empirically observed prices at the day-ahead and balancing power market, January–April 2006, with original Erev and Roth reinforcement learning, proportional action selection, recency $\phi = 0.1$, experimentation $\varepsilon = 0.2$	153
B.5	Simulated and empirically observed prices at the day-ahead and balancing power market, May–August 2006, with original Erev and Roth reinforcement learning, proportional action selection, recency $\phi = 0.1$, experimentation $\varepsilon = 0.2$	154
B.6	Simulated and empirically observed prices at the day-ahead and balancing power market, September–December 2006, with original Erev and Roth reinforcement learning, proportional action selection, recency $\phi = 0.1$, experimentation $\varepsilon = 0.2$	155

B.7 Simulated and empirically observed prices at the day-ahead and balancing power market, January–April 2006, with modified Erev and Roth reinforcement learning, proportional action selection, recency $\phi = 0.1$, experimentation $\varepsilon = 0.2$ 156

B.8 Simulated and empirically observed prices at the day-ahead and balancing power market, May–August 2006, with modified Erev and Roth reinforcement learning, proportional action selection, recency $\phi = 0.1$, experimentation $\varepsilon = 0.2$ 157

B.9 Simulated and empirically observed prices at the day-ahead and balancing power market, September–December 2006, with modified Erev and Roth reinforcement learning, proportional action selection, recency $\phi = 0.1$, experimentation $\varepsilon = 0.2$ 158

B.10 90% confidence intervals for $E(Z_j)$, comparison of neighboring scenarios of $p^{DAM,max}$ (action domains; Q-learning simulations).... 159

B.11 90% confidence intervals for $E(Z_j)$, comparison of different values of portfolio integration parameter ψ with reference value (Q-learning simulations)..................................... 160

B.12 90% confidence intervals for $E(Z_j)$, comparison of scenario without emissions trading with reference scenario (Q-learning simulations) 160

C.1 90% confidence intervals for $E(Z_j)$, comparison of lowest and highest scenarios of tendered balancing power capacity (Q-learning simulations)..................................... 161

C.2 90% confidence intervals for $E(Z_j)$, comparison of lowest and highest scenarios of tendered balancing power capacity (original Erev and Roth RL simulations) 161

C.3 90% confidence intervals for $E(Z_j)$, comparison of lowest and highest scenarios of tendered balancing power capacity (modified Erev and Roth RL simulations) 162

C.4 90% confidence intervals for $E(Z_j)$, comparison of pay-as-bid and uniform pricing (Q-learning simulations) 162

C.5 90% confidence intervals for $E(Z_j)$, comparison of pay-as-bid and uniform pricing (original Erev and Roth RL simulations) 162

C.6 90% confidence intervals for $E(Z_j)$, comparison of pay-as-bid and uniform pricing (modified Erev and Roth RL simulations) 163

C.7 90% confidence intervals for $E(Z_j)$, comparison of DIV4,0.50 scenario with reference scenario (Q-learning simulations) 163

C.8 90% confidence intervals for $E(Z_j)$, comparison of DIV4,0.50 scenario with reference scenario (original Erev and Roth RL simulations) 163

C.9 90% confidence intervals for $E(Z_j)$, comparison of DIV4,0.50 scenario with reference scenario (modified Erev and Roth RL simulations) 164

C.10 90% confidence intervals for $E(Z_j)$, comparison of DIV8,0.50 scenario with reference scenario (Q-learning simulations) 164

C.11 90% confidence intervals for $E(Z_j)$, comparison of DIV8,0.50 scenario with reference scenario (original Erev and Roth RL simulations) 164

C.12 90% confidence intervals for $E(Z_j)$, comparison of DIV8,0.50 scenario with reference scenario (modified Erev and Roth RL simulations) 165

List of Tables

2.1	Procured balancing power quantities of German TSOs in 2006	13
3.1	Summarized overview of agent-based electricity market modeling approaches	50
4.1	Agents and power plants in the simplified electricity market scenario	70
4.2	Tested variants and parameter values of regarded learning algorithms	73
4.3	Average market prices of last 200 iterations for simulations with varied action domains	77
4.4	Appropriate variants of learning algorithms	82
6.1	Divestiture scenarios	132
A.1	Summarized overview of agent-based electricity market modeling approaches	144
B.1	UCTE load for 2006 in MW	147
B.2	Generator agents and power plant characteristics of the reference scenario	148
B.3	Emissions scenario: generating agents	149
B.4	Emissions scenario: other CO_2 emissions trading participants	149

Part I
Motivation and Fundamentals

Chapter 1
Introduction

Electricity markets rank among the most complex of all marketplaces operated at present. The institutions that set the framework for power generation, transportation and retail delivery must be designed in a way to ensure that the three main goals of energy supply can be met: environmental sustainability, economic efficiency and security of supply. Simultaneously, technical reliability requirements, as well as transmission and unit commitment constraints must be taken into account. These requirements have resulted in the development of several interrelated markets, such as power exchanges, balancing power markets, and markets for emission allowances, among others. The ensemble of these interrelated markets has to be analyzed together if advice for good market design is to be derived. The analysis of markets in the electricity sector is further complicated by the presence of few dominating and vertically integrated firms. Models that assume perfect competition are inappropriate for representing this oligopoly of bidders in a realistic manner. While many theoretical approaches assume that market participants will bid competitively, and only interact once, wholesale electricity markets involve an oligopoly of bidders who trade repeatedly. This constellation encourages strategic bidding. Players seek to ameliorate their bidding strategies continually, based on their experience gained in previous trading days. Thus, realistic electricity modeling must account for strategic bidding by dominant players, learning from daily repeated interaction, and the dynamics stemming from multiple interrelated markets. In this context, the question arises which market rules or which regulatory framework is most appropriate to ensure efficient market outcomes.

Following the postulation formulated by Roth (2002), Marks (2006), or Weinhardt, Holtmann, and Neumann (2003), among others, markets should be designed using engineering methods, such as experimentation and computation. Roth (2002) identifies three types of markets which have been subject to design efforts by economists and which require appropriate engineering tools in order to harness the markets' complexities. Electricity markets belong to this category; the other markets mentioned are labor clearinghouses and radio frequency auctions. Besides, auctions on the Internet are noted as a new opportunity to investigate the effects of different auction designs. Similarly, Marks (2006) gives five examples for "designed markets" that have to be carefully engineered by economists. Besides electricity markets, Marks also mentions markets for emission rights, such as CO_2 allowances; the other three are stock markets, auctions for electro-magnetic spectrum, and on-line markets in the e-commerce field. Weinhardt et al. (2003) also call for an engineering approach to designing electronic markets; they describe a more holistic procedure that does not only take into account the *market design* perspective, which identifies desirable market micro structures, but also encompasses the *infrastructure* and *business structure* of an electronic market, defining how market participants can communicate and how the service offered through the marketplace should be priced.

The complexity of the electricity sector and its high importance for a competitive economy calls for modeling methods that help gaining insights into the dynamics of power markets, and that are capable of properly representing the relevant complex aspects. Computational methods, and agent-based (AB) simulations in particular, are useful tools that support analysis in the process of engineering complex markets. They also offer the possibility of modeling agents that adapt to their environment, i.e. that *learn* their best strategies from repeated interaction, thereby following another proposition by Roth (2002): "models of learning with the ability to predict behavior in new environments will be a valuable addition to the designer's toolbox [...]". Agent-based modeling offers great flexibility for specifying different scenarios and may help overcome some of the shortcomings of traditional methods. AB simulation models can be used as fully controllable virtual laboratories for testing economic design alternatives in order to determine the particular market designs that perform best in an environment of selfish agents (Tesfatsion, 2006). Consequently, in pursuing the goal of engineering electricity markets, this work relies on agent-based modeling as its main methodology.

1.1 Objectives and Research Questions

The research work at hand attempts to contribute to the endeavor delineated here in two ways: first, it adds to the development of the AB methodology for electricity market simulation and second, it develops a model of the German electricity system that may serve as a tool for engineering electricity markets.

The aim of the research conducted in this thesis is to implement an agent-based simulation model that realistically represents those parts of the power sector that are relevant to the analysis of daily wholesale electricity trading and CO_2 emissions trading. Market participants are represented as adaptive software agents facing the problem of trading on several interrelated markets, i.e. a day-ahead electricity market, a market for balancing power (positive minute reserve) and a market for CO_2 emission allowances. They develop trading strategies through different learning algorithms. In order to ensure robustness of simulation results, several learning models will be applied and results are compared. In the model, markets will be interrelated only through the agents' trading strategies. When searching for profit maximizing bidding strategies, agents consider opportunity costs, i.e.

foregone profits that they could have realized on the other markets. Through this procedure, they coordinate their biding behavior on all three simulated markets.

The model will be applied to the case of the German electricity industry, taking data of power plants and electric load as inputs. From a methodological perspective, the research question that has to be answered first is whether an agent-based model is capable of realistically representing the German electricity sector with strategically interacting agents in interrelated markets. It will be tested whether results obtained from the simulations are robust and can be considered valid. Therefore, simulation outcomes are compared to empirically observed market prices of the year 2006. Within the challenge of engineering electricity markets, the questions to be answered in the work at hand are the following:

1. Which are the main factors influencing price formation in interrelated electricity and emissions markets?
2. Which market rules lead to more competitive electricity market outcomes?
3. Which market structures are most vulnerable to market power exertion by the agents?
4. What are the implications of different conceivable policy measures that alter the framework of the electricity industry?

1.2 Thesis Structure

This thesis is structured into three parts which will be briefly described in the following. For a quick reader's guide, a graphical presentation of the thesis structure is provided in Fig. 1.1.

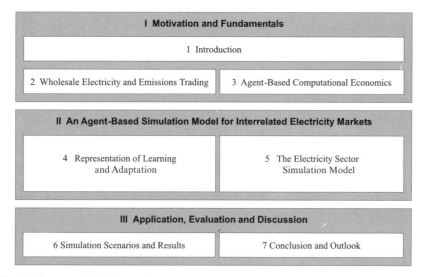

Fig. 1.1 Structure of the dissertation

Part I – Motivation and Fundamentals introduces the research questions and explains the most important concepts necessary for understanding the chosen approach. Chapter 2 describes the German electricity system's characteristics, including a discussion of the interrelations between several important short-term markets in the sector. The methodology of Agent-Based Computational Economics[1] is introduced in Chap. 3. This chapter also includes a critical discussion of related work in the field of agent-based simulations in wholesale electricity market modeling.

Part II – An Agent-Based Simulation Model for Interrelated Electricity Markets delivers a detailed presentation of the model developed in this work. Chapter 4 is devoted to an in-depth discussion of possible representations of the agents' adaptive behavior. It is followed by a description of the model characteristics, comprising the overall architecture, the modeled agent behavior, and implementation details (Chap. 5). This chapter also contains a section that describes the empirical validation procedure and a detailed sensitivity analysis.

Part III – Application, Evaluation and Discussion illustrates all simulated scenarios that are run for the analysis of policy questions relevant in the German electricity sector, and presents an evaluation and discussion of the results obtained from these simulations (Chap. 6). The thesis concludes with a summary of the main results and an outlook on future research in Chap. 7.

Parts of the work presented here have already been published in other articles, which are listed in the following: Weidlich and Veit (2006), Veit, Weidlich, Yao, and Oren (2006), and Weidlich and Veit (2008a–d).

---

[1] Throughout this work, the terms "Agent-Based Computational Economics", "agent-based (computational) modeling", or "agent-based simulations" are used synonymously.

Chapter 2
Wholesale Electricity and Emissions Trading

The European electricity sector liberalization, introduced through the Directive 96/02/EG of the European Union (1996), entailed a separation of the main parts of the electricity value chain: generation, transmission, distribution and retail supply.[1] In Germany, the Energy Industry Act (Energiewirtschaftsgesetz, EnWG, 2005) regulates the electricity sector and defines the rules of operation and competition for all firms and organizations engaged in power generation, transportation and supply.

Although generation, transmission, distribution and retail supply have been separated, the incumbent firms operating in the German electricity industry are active in several or all of these fields, i.e. they are vertically integrated. The European and German legislation enforces that distribution and transmission activities are separated organizationally from generation and retail supply, and that non-discriminating access to the grid is allowed to any power generator or retail supplier. In Germany, the four *transmission system operators* (TSOs) are vertically integrated with power generation and retail firms belonging to the same trust. At the same time, these four large companies are the dominant players in power generation; together, they operate approximately 80% of the total installed generation capacity. Thus, the market can be characterized as an oligopoly and the large four companies have some potential to act strategically.

The generation and retail supply parts of the electricity value chain are subject to competition. On the generation side, power plant operators (*generators*) compete for selling power to the wholesale markets. On the retail supply side, *load serving entities* (LSEs) buy on the wholesale markets the electric power that they need to serve their retail customers. One central marketplace where trades between generators, load serving entities and intermediaries are settled is the power exchange. Exchanges usually also offer derivative products that help market participants hedge their risks. These marketplaces are outlined in Sect. 2.1.1.

---

[1] Transmission is the bulk transfer of electric power through extra high voltage lines (220 or 380 kV). Distribution is the transport of electricity within regional power grids of high voltage (110 kV), medium voltage (around 10–20 kV) and low voltage (230 or 400 V) lines; these lines transport electricity from the transmission grid to the locations of end use.

The power grids, i.e. the transmission and distribution systems, are regarded as natural monopolies. They are operated by transmission and distribution system operators, respectively, who are regulated and monitored by a responsible governmental authority. Where possible, competitive elements are introduced into this regulation. This is the case, for example, with the procurement of reserve capacity for frequency control in the transmission system. Reserve capacities are traded on balancing power markets, which are described in Sect. 2.1.2.

Another European legislation has a considerable impact on competition in electricity generation. Through the establishment of a greenhouse gas emission allowance trading scheme (Directive 2003/87/EC of the European Union, 2003), generators are enforced to hold allowances for every ton of carbon dioxide (CO_2) they emit through power production. By means of this legislation, CO_2 emissions become one important cost factor of electricity generation, and daily trading strategies have to consider allowance prices as well. The European emissions trading scheme and its application in Germany are surveyed in Sect. 2.2.

A generator who owns power plants that are qualified to deliver minute reserve has to decide how to offer this capacity on day-ahead electricity markets and on the balancing power market in a profit-maximizing manner. Besides, generators have to take into account opportunity costs that arise from the possibility to trade CO_2 emission allowances. As these aspects play an important role in the model that is developed in the second part of this work, these interrelations are discussed in Sect. 2.3.

2.1 Electricity Marketplaces

Different marketplaces for trading electric power, reserve capacity or derivative products have emerged in restructured electricity systems. In the following, power exchanges and balancing power markets are described in more detail, and the peculiarities of these markets in Germany are highlighted.

2.1.1 Power Exchanges

One important development after electricity sector liberalization in Europe was the establishment of power exchanges. In contrast to the traditional means of electricity trading, that is mainly bilateral *over the counter* (OTC) contracts, exchanges offer marketplaces at which standardized electricity products can be traded. Centralized trading at power exchanges offers the advantage of high liquidity, and it delivers valuable price information to the whole sector. The standardization of electricity products makes trading easier, thus lowering transaction costs; at the same time, comparability of prices increases. Within the last years, most power exchanges have diversified their service portfolios through offering additional markets for electricity-related commodities such as coal and gas futures, or CO_2 emission allowances.

2.1 Electricity Marketplaces

There are several possible ways of trading electricity at exchanges. The dominant market institution for electricity trading, however, is the double-auction format. In a sealed bid double-auction, both buyers and sellers submit bids specifying the prices at which they are willing to buy or sell a certain good. Buying bids are then ranked from the highest to the lowest, selling bids from the lowest to the highest bid price. The intersection of the so formed stepwise supply and demand functions determines the market clearing quantity and gives a range of possible prices from which the market clearing price is chosen according to some arbitrary rule (McAfee & McMillan, 1987). Double-auctions deliver efficient allocations if the number of sellers and buyers is sufficiently large (Wilson, 1985). Sealed bid double-auctions that are executed at discrete points in time are also called *call markets*, in contrast to *continuous double-auctions*. In this latter case, an order book is updated each time a new bid is submitted; executability is checked immediately, and prices are determined individually for each executed trade. In the case of electricity trading, various amounts of electric energy can be traded in one auction. Electricity auctions are therefore characterized as *multi-unit auctions*.

In the German electricity sector, the European Energy Exchange AG (EEX), which is situated in Leipzig, operates the main power exchange. The traded volume at the EEX spot market accounted for 88.7 TWh in 2006. This corresponds to more than 16% of the total amount of electricity consumed in Germany in the same year. With these high trading volumes, the European Energy Exchange is one of the largest and most important power exchanges in Europe; clearing prices resulting at the EEX markets are also used as reference prices for other electricity contracts in Germany and elsewhere in Europe. Other important power exchanges in Europe are the Scandinavian Nordpool, the Amsterdam Power Exchange, or the French Powernext.

Day-Ahead Trading

One important marketplace operated by power exchanges is the market for contracts that entail physical electricity delivery at a specified time on the following day. These kinds of markets are referred to as *spot markets* or – more accurately – *day-ahead markets*.[2]

The subsidiary company EEX Power Spot GmbH operates a day-ahead market for hourly and block electricity contracts. This marketplace offers both continuous trading and a call market (EEX, 2007). At the call market, contracts for every single hour of the following day, as well as block bids of several consecutive hours can be traded. Bids for block contracts are integrated into the auctions for single hour contracts through an iterative procedure that makes sure that all hours of a specific

---

[2] The term *spot market* is misleading in this context; usually, spot markets are understood as marketplaces where contracts for standardized goods are traded and immediately fulfilled. As in practice electricity delivery is also traded at a shorter term than one day ahead (e.g. intra-day trading, see below in this section), the kind of markets where trades are physically settled on the day subsequent to trading should more accurately be termed *day-ahead markets*.

block are either executed jointly or not executed at all. At the continuous auction, only pre-defined block contracts can be traded. The EEX trading system for hour and block contracts is based on the Xetra® electronic trading system.

The EEX day-ahead market distinguishes several places of delivery, which correspond to the balancing zones of the four German TSOs, and the Austrian and Swiss TSO.[3] If no transmission constraints are present, electricity prices for each hour are the same in the whole market area, which comprises the German and Austrian market (the Swiss market constitutes an own market area). Only in cases in which congestion occurs between single balancing zones, different prices are calculated for each zone. However, this situation is extremely rare because there is sufficient transmission capacity between the zones.

Derivatives and Intra-Day Trading

Electricity markets are characterized by high price volatility. Major causes of those volatilities are demand and supply fluctuations due to changing weather conditions, availability of generation capacity, and fuel costs or other production costs. Electricity futures and other derivatives, such as options, swaps, or forward contracts, help generators and load serving entities manage their price risks arising from volatility. Futures and forwards are the most important electricity derivatives. Futures are legally binding and negotiable contracts that call for the future delivery of electricity at a specified date, location and quantity. In most cases, physical delivery is not effected directly by the seller of the futures, but the futures contract is closed by buying or selling a delivery contract on or near the delivery date. A forward contract also defines an obligation to deliver electricity at a specified date, location and quantity and at a fixed price. In contrast to futures, the terms and conditions of forward contracts are not standardized, but negotiated to meet the particular financial or risk management needs of the parties involved (Stoft et al., 1998). For this reason, they are not traded at exchanges, but usually through bilateral trade or through specialized intermediaries.

At the EEX, futures based on market price indices are available for current and future months, quarters, and years. Other major exchanges in Europe also offer standardized futures contracts with electricity price indices as underlying. These are usually traded in continuous double-auction markets.

On the very short term, generators may want to sell spare capacity or purchase additional electric energy in order to be able to react to the actual supply and demand situation after closure of the day-ahead market. Contracts of electricity delivery a few hours ahead of physical fulfillment have usually been negotiated *over the counter*. Since September 2006, the EEX intra-day market allows trades in power contracts for delivery on the same or following day. These can be accomplished until 75 min before physical delivery. Intra-day trading takes the form of a continuous double-auction with an open order book where anonymous buying and selling bids can be entered at any time.

---

[3] The balancing mechanism and the role of the TSO are introduced in Sect. 2.1.2.

2.1.2 Balancing Power Markets

Reliable and secure power grid operation is a necessary prerequisite for wholesale electricity trading. Transmission system failures or instable operation not only disturb the affected balancing zone, but also interfere with networks in other regions, as the German transmission systems are coupled to those of neighboring countries within the framework of the *Union for the Co-ordination of Transmission of Electricity* (UCTE). In order to ensure stable operation, the UCTE issues security and reliability standards that have to be followed by all interconnected TSOs (UCTE, 2006). In addition, the *Verband der Netzbetreiber e.V.* (VDN, 2007) specifies the rules according to which the TSOs in Germany have to operate their transmission grids.

There are four balancing zones in the German electricity system which are operated by the four TSOs E.ON Netz GmbH, RWE Transportnetz Strom GmbH, Vattenfall Europe Transmission GmbH and EnBW Transportnetze AG. The TSOs ensure stable grid operation at the extra high voltage level. Processing of electricity deliveries or trading transactions through the transmission lines is effected within *balance agreements* between the TSO and the trader. Every trader who wants to buy or sell electricity at the wholesale level in one of the four balancing zones has to have a *balancing area* within this zone. The person in charge for the balance area has to ensure that the sum of power feed-in into the balance area equals the sum of power withdrawal from it throughout all 15-min periods.

In practical operation, it is almost inevitable that imbalances between feed-in and withdrawal occur due to stochastic demand, unpredictable plant outages, or fluctuating renewable energy production. In order to ensure a stable operation of the transmission grid, the TSOs continually level out imbalances that occur in the balancing areas belonging to their balancing zone. The process of balancing the transmission grid (frequency control) is described in the following.

Frequency Control

In any power system, the electric energy generated must be maintained at a constant equilibrium with the energy consumed. Disturbances in this balance cause a deviation of the system's frequency, which must be avoided in order to prevent damage of the consumption devices connected to the power grid. There is only very limited possibility of storing electric energy. Consequently, the production system has to be sufficiently flexible in changing its generation level in order to maintain the power equilibrium in real-time. It must be able to handle both changes in power demand and outages in generation and transmission instantly.

Frequency control is effected in a three-stage process which is divided into primary, secondary and tertiary control. The balancing power qualities of the three stages differ in terms of required activation and response speed. In each of the stages the TSO has to provide a certain amount of incremental and decremental capacity, which is held in reserve for situations in which, respectively, negative or positive deviations from the power balance occur. The frequency control process is depicted in Fig. 2.1.

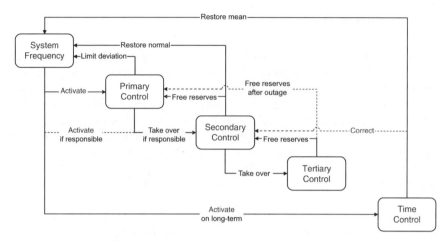

Fig. 2.1 Frequency control as instructed by UCTE (2006)

Demand increases or plant outages will cause the system frequency to decrease; reversely, demand decreases cause the frequency to increase. In both cases, regulating units will perform automatic *primary control* in order to re-establish the balance between electricity demand and supply and to keep the frequency near its set-point of 50 Hz. The UCTE stipulates that in normal operation, a sudden failure of 3,000 MW of the generating capacity, without other disruptions, has to be corrected solely by the action of the primary controller and must not require load-shedding.

Secondary control will take over load balancing for the remaining frequency and power deviation after 15–30 s. The function of secondary control is to restore the power balance and, consequently, to keep or to restore the system frequency to its set-point value of 50 Hz. The deployment of secondary control reserve ensures that the full reserve of primary control power activated will be made available again. These actions of secondary control will take place simultaneously and continually, both in response to minor deviations (which will inevitably occur in the course of normal operation) and in response to a major discrepancy between production and consumption. While primary control needs to be available within a few seconds (up to 30 s at the latest), activation of secondary reserve may take several minutes time, with a maximum of 5 min.

Tertiary control reserve, also denoted as *minute reserve*, is automatically or manually activated in order to guarantee the provision of an adequate secondary control reserve at the right time, or distribute the secondary control power to the various generators in the economically best possible way. Minute reserve usually comes into action when larger frequency imbalances occur. The time-frame in which minute reserve is scheduled is typically 15 min.

When consumption exceeds production on a continuous basis, immediate action must be taken to restore the balance between the two (by the use of standby supplies, contractual load variation or load-shedding or the shedding of a proportion of customer load as a last resort). Sufficient transmission capacity must be maintained at all times to accommodate reserve control capacity and standby supplies.

2.1 Electricity Marketplaces

Procurement Auctions for Balancing Power

The generation capacity needed for primary, secondary and tertiary frequency control are procured separately by means of procurement auctions. The costs arising from primary control and from holding secondary and tertiary control capacity in reserve are constituent of the transmission system usage fee. Costs for the deployment of electric work from secondary control or minute reserve plants are charged to the account of the balance areas that caused imbalances. The legal framework for these procedures of balancing power procurement and billing are set by the Energy Industry Act (EnWG, 2005), and by two other more specific provisions of law: *Stromnetzentgeltverordnung* (StromNEV, 2005) and *Stromnetzzugangsverordnung* (StromNZV, 2006).

In the course of the years 2001 and 2002, the four German TSOs each established Internet-based balancing power markets through which they procure the three types of incremental and decremental balancing power by way of auctions. Only pre-qualified bidders can take part in the balancing power procurement auctions; the pre-qualification process ensures that the participating power plants meet the technical requirements that are defined for each of the balance reserve qualities. Primary and secondary reserve capacity is procured twice per year for six-month periods, whereas minute reserve auctions take place on a daily basis. The quantities of primary, secondary and tertiary control reserves that have been tendered by the TSOs in 2006 are summarized in Table 2.1. Since December 2006, the four TSOs are required to procure their balancing power demand in a joint Internet-based auction.[4] In her latest adjudication, the responsible regulatory authority has refined the specifications of these auctions to a more precise level (Bundesnetzagentur, 2006). These specifications define – among other things – that the daily procurement auctions for minute reserve have to take place before spot market clearing, in order to allow for greater liquidity on the balancing power markets.

Table 2.1 Procured balancing power quantities of German TSOs in 2006

	Primary reserve (MW)	Secondary reserve (MW)	Minute reserve[a] (MW)
E.ON Netz GmbH	$\pm\|163\text{–}164\|$	$+800$ -400	$+1,100$ -400
RWE Transportnetz Strom GmbH	± 285	$+1,230$ $-\|980\text{–}1,230\|$	$+\|930\text{–}1,080\|$ $-\|760\text{–}810\|$
Vattenfall Europe Transmission GmbH	$\pm\|137\text{–}140\|$	$+580$ -580	$+730$ -530
EnBW Transportnetze AG	$\pm\|71\text{–}73\|$	$+720$ -390	$+\|390\text{–}510\|$ $-\|300\text{–}330\|$

[a] January–November 2006. The procurement procedure has been altered on 1 December 2006

---

[4] At the beginning, this legal requirement has only been fulfilled for minute reserve auctions; joint auctions for primary and secondary reserve capacity have started in December 2007.

Balancing power markets can be characterized as single-sided, multi-unit procurement auctions. Primary reserve bids consist of the offered quantity (capacity) and of the price asked for electric work (energy) that is actually deployed. In contrast, offer bids for secondary and minute reserve contain two bid prices: the capacity price and the energy price. Capacity prices are paid for holding the specified generation capacity in reserve (and, thus, loosing the opportunity to sell its output elsewhere), and energy prices are paid for the electric work that is actually deployed for balancing purposes. Several mechanisms are conceivable for selecting successful bids based on the two bid prices. The difficulty with optimization procedures that minimize the expected cost for the buyer of balancing reserve (i.e. the TSO) is that the actual reserve deployment is unknown at the time when bids are selected. In practice – in the case of minute reserve procurement – the decision of selecting successful bids is based solely on the capacity price. At the time of balancing deployment, a *merit order* of energy prices of all successful bids is formed, that is bids are ranged from the lowest to the highest energy bid price. If positive or negative balancing power is needed, the decision which capacity to use is based on this merit order. The described procedure is transparent and simple, but leaves bidders with some leeway for gaming the mechanism, e.g. through bidding strategic energy prices (Bundesnetzagentur, 2006).

Besides bid selection, another design criterion of the balancing power market mechanism is how to remunerate successful bids. Two options are discussed in the literature and are employed in practice. These are the *discriminatory*, or *pay-as-bid* and the *uniform price* settlement. In the former case, every successful bid is paid its bid price, whereas in the latter case, one market clearing price is determined – usually the price of the highest successful bid – which is paid to all successful bidders. The German legislation has opted for the pay-as-bid settlement scheme (Bundesnetzagentur, 2006) for minute reserve markets.

2.2 Emissions Trading

The European emissions trading scheme (EU-ETS, European Union, 2003) has been introduced in order to stimulate a development towards more efficient and less carbon-intensive means of production, especially in the electricity sector. A *cap* limits the total amount of greenhouse gases[5] that the industries covered by the EU-ETS are allowed to emit. Trading emission allowances permits to realize target emission reductions at the lowest overall cost (Dales, 1968).

---

[5] Usually, the discussion about emissions trading is centered around emissions of carbon dioxide (CO_2), though the Kyoto protocol also covers five other greenhouse gases, i.e. methane (CH_4), nitrous oxide (N_2O), perfluorocarbons (PFCs), hydrofluorocarbons (HFCs), and sulphur hexafluoride (SF_6). Their ability to trap heat in the atmosphere (thus contributing to global warming) is expressed through the *Global Warming Potential*, which is calculated in relation to that of CO_2. The first National Allocation Plans only covered CO_2 emissions. In the following, solely CO_2 will be regarded, as it accounts for the largest share of greenhouse gas emissions in the electricity industry.

2.2 Emissions Trading

While the Directive establishing the EU-ETS sets the framework of the trading system's functioning, some design issues have to be fixed by the Member States. They have to specify the total amount of allowances they want to issue, and need to define how these allowances are allocated to the plants participating in the emissions trading scheme. These rules are specified in the States' *National Allocation Plans* (NAPs).

Two basic methods of allowance allocation to existing plants are possible. They might either be *auctioned* among the emitters, or they can be allocated free of charge, according to a fixed allocation method. Allocation rules can be based on past emissions in a selected baseline period (*grandfathering*), or on the output in a recent reference period and an overall target, expressed in CO_2 per unit of output (*benchmarking*). The method of allowance allocation to existing installations does not basically affect the static efficiency of emissions trading. Overall investment patterns under an emissions trading system depend mainly on the stringency of the overall emission cap and, thus, on the resulting price of allowances. However, the allocation method affects the competitive situation of companies. Some companies that would be net sellers under grandfathering could become net buyers under benchmarking. Besides, many researchers (e.g. Diekmann & Schleich, 2006; Jung et al., 1996; Milliman & Prince, 1989) argue that auctioned allowances would create greater incentives for technology diffusion and adoption than allowances allocated free of charge, since technology diffusion reduces the price of allowances. The innovator can benefit from this price decrease, since he will not have to pay as much for the rest of his emissions. In the case of free allocation, however, the price decrease due to innovation would lower the value of the innovator's allowances, which makes innovation less attractive. The EU-ETS opted for gratuitous distribution of the majority of emission allowances. Auctioning is only permissible for 5% of allowances in the pilot phase (2005–2007) and 10% in the second period of European emissions trading (2008–2012). The German NAPs for the first (BMU, 2004) and second trading period (BMU, 2007) did not take advantage of this option and allocate all allowances gratuitously, based on a grandfathering procedure.

Other design options for emission trading are the treatment of new entrants and plant closure. Emission allowances might be issued free of charge to new installations; in the opposite case, operators of new power plants would have to buy all allowances needed for generation on the market. Correspondingly, allowances allocated free of charge might expire if the plant which they are allocated to is shut down during the commitment period. Alternatively, allowances might retain their value for a certain time after plant closure, at the most until the end of the commitment period. The effects of these rules will not be discussed here; the interested reader is referred to, e.g. Graichen and Requate (2005).

Figures 2.2 and 2.3 show the price developments for emission allowance spot contracts and for futures with different expiration dates. Settlement prices at the EEX and at the European Climate Exchange (ECX)[6] are depicted. It can be seen that the prices for allowances of the first EU-ETS period have experienced a sharp

---

[6] The European Climate Exchange is a marketplace based in London and Amsterdam which is specialized on emission allowance trading (http://www.europeanclimateexchange.com/).

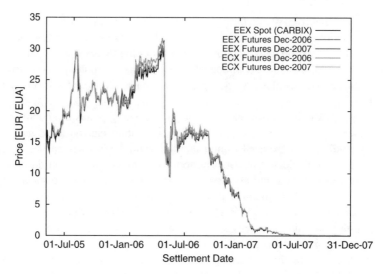

Fig. 2.2 Settlement prices for CO_2 futures and spot contracts of the first EU-ETS period (2005–2007)

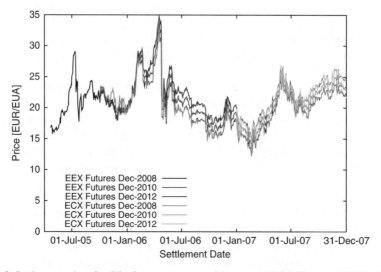

Fig. 2.3 Settlement prices for CO_2 futures contracts of the second EU-ETS period (2008–2012)

drop and have been traded at prices below ~ 0.10 EUR/EUA[7] since mid-June 2007. This development is not surprising, as the number of allowances issued for the first EU-ETS period was very generous by all Member States.

---

[7] One EU Allowance (EUA) allows for the emission of 1 ton of CO_2.

In the second trading period, which starts at the same time as the emissions trading scheme installed through the Kyoto protocol in 2008, allowance distribution is tighter; the EU has rejected many NAPs for the second period because the emission cap fixed in the plans were too high. Given stricter emission targets, prices can be expected to be higher in the second trading period. This expectation is also reflected in the prices of EUA futures with expiration dates of 2008 and beyond (see Fig. 2.3). Between May 2006 and December 2007, futures for allowances of the second EU-ETS period have been traded at prices between 12 and 27 EUR/EUA.

Through the obligation to hold allowances for every ton of carbon dioxide emitted, the latter have become one important cost factor in power generation. A small sample calculation should illustrate this: the CO_2 emission factor of hard coal is 338 kg/MWh, so a coal fired power plant with an efficiency of 45% would emit around 750 kg/MWh. With an allowance price of 20 EUR/EUA the additional generation cost caused by CO_2 emissions is near 15 EUR/MWh. At the same allowance price, a natural gas fired combined cycle power plant with an efficiency of 57% and a fuel emission factor of 202 kg/MWh would incur around 7 EUR/MWh of CO_2 emission costs. Compared to average hourly electricity spot prices, which were around 45 EUR/MWh between 2005 and 2007, this cost factor is not negligible.

2.3 Interrelations Between Day-Ahead, Balancing and Allowance Markets

Operators of electricity generation assets that are able to deliver balancing power have to decide on which market to bid their capacity. Once they have committed (part of) their capacity on one market, they cannot sell the same capacity on another market. Generators seek to maximize profits under consideration of the different option at hand and, thus, face opportunity costs. These are the costs of an action in terms of foregone opportunities, i.e. benefits that could have been realized from the (most valuable) action not taken. In daily electricity market bidding, selling capacity on the day-ahead power exchange entails the opportunity cost of foregone profits on the market for incremental minute reserve and vice versa. A rational bidder would thus integrate the opportunity cost of one market into the bid on the other market. Prices for minute reserve depend to a considerable extent on these kinds of opportunity costs.

Interrelations between the day-ahead electricity market and auctions for decremental minute reserve are less obvious. Delivery of decremental minute reserve is only possible if output from the power plant has already been sold, i.e. the plant will be running on the following day (Swider, 2006). A plant operator who has already sold decremental minute reserve will thus have to make sure to sell his capacity elsewhere and might bid at lower prices on the day-ahead market in order to increase his chance of being successful there. Primary and secondary reserves are auctioned for half-year periods, so these markets are more related to longer term delivery contracts and futures contracts.

Further opportunity costs in electricity trading arise from emissions trading. A plant operator who owns CO_2 emission allowances can decide to either use his allowances for obtaining the right to emit the CO_2 caused by his electricity output, or to sell allowances and thereby gaining a certain profit. In order to optimize the overall trading strategy, opportunity costs of selling allowances have to be taken into account when bidding on electricity markets. In case the plant operator does not own (sufficient) emission allowances, he has to purchase the required allowances for the emissions incurred from electricity generation on the market. In this case, emission allowances constitute an effective cost factor, which has to be taken into consideration as well.

As Germany's electricity production relies to a great extent on the CO_2 intensive fuels hard coal (21% of all electricity produced in Germany) and lignite (24%; data of 2006, BMWI, 2007), it can be expected that prices for emission allowances also influence electricity prices. Figure 2.4 depicts day-ahead electricity and emission allowance prices, both from the EEX spot market. It can be seen that daily electricity prices do not correlate strongly with allowance prices. In fact, the Pearson's correlation coefficient between the two variables is 0.31 for daily average prices on workdays.[8] This correlation can be considered weak, but it is stronger than in countries with a less CO_2 intensive electricity production (the correlation coefficient between weekly electricity and allowance prices in Austria, for instance, where hydroelectric generation is predominant, is 0.22 according to Kemfert et al., 2007). However, on a longer-term time horizon, a considerable impact of emissions trading on electricity prices can be observed. An analysis of half-year average prices reveals that prices have risen considerably after the introduction of the EU-ETS, and the Pearson's correlation coefficient between half-yearly averages of electricity and allowance prices from 2005 to 2007 is 0.50[9] (see right side of Fig. 2.4; note that the scale for electricity prices differs from that of the left side figure). Even though other aspects,

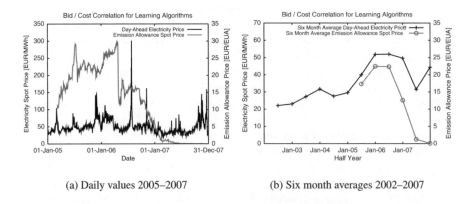

(a) Daily values 2005–2007   (b) Six month averages 2002–2007

Fig. 2.4 Day-ahead electricity and emission allowance spot settlement prices

---

[8] Emission allowances are not traded on weekdays and national holidays.

[9] The Pearson's correlation has been even higher (0.84) between January 2005 and June 2007.

like fuel prices, surely had an impact on prices in the years of 2005–2007 as well, the strong correlation between allowance and electricity prices and the temporal coincidence of rise in electricity prices and the introduction of the EU-ETS makes it plausible that emissions trading caused an important part of the price increase.

Prices for other commodities such as coal or natural gas and derivatives such as electricity futures also influence trading strategies on markets for day-ahead electricity contracts, balancing power and emission allowances. However, the focus of this work is placed on daily bidding on the three specified markets; thus, interrelations with other markets are not further discussed here.

2.4 Summary

This chapter described the main characteristics of wholesale power markets and of CO_2 emissions trading, and shed light on the specifications of these markets in the German electricity industry. It highlights the fact that there is not only one power market, but electricity and related products are traded on several markets that are interrelated with each other. As there are a few powerful actors, the electricity industry is not structured in a perfectly competitive manner. It must rather be qualified as an oligopoly. In addition with the daily repeated interaction, players have some potential to learn bidding strategies that deviate from those that can be expected in perfect competition. Moreover, trading strategies must be analyzed by taking into account all options that a power plant operator has through the opportunity costs that arise from these options, including bidding on day-ahead electricity markets, balancing power markets and markets for CO_2 emission allowances.

Chapter 3
Agent-Based Computational Economics

As outlined in Chap. 2 the electricity sector is characterized by technical constraints, multiple interlinked markets, and an oligopolistic structure with vertical integration. These aspects make electricity markets rank among the most complex of all markets, and push most classical modeling methods to their limits (for a discussion of prevalent electricity market modeling methods, the reader is referred to Ventosa et al., 2005). Equilibrium models do not consider strategic bidding behavior or assume that players have all relevant information about the other players' characteristics and behavior; they also disregard the consequences of learning effects from daily repeated interaction (Rothkopf, 1999). Game theoretical analysis yields insights into specific aspects of electricity trading (Wilson, 2002), but is usually limited to stylized trading situations among few actors, and places rigid – oftentimes unrealistic – assumptions on the players' behavior. Human subject experiments are difficult to apply to electricity market research, as some expertise is necessary to realistically imitate the bidding behavior of a power generator. Hence, experiments are deemed appropriate for simple electricity trading scenarios only.

The complexity of the electricity sector and its high importance for a competitive economy calls for new modeling methods that facilitate gaining insights into various aspects of power markets. Agent-based (AB) modeling is one appealing new methodology that has the potential to overcome some of the aforementioned shortcomings of optimization or equilibrium modeling methods. Within the last ten years, more and more researchers have developed electricity market models with adaptive software agents. This field of research is still growing and maturing.

In the following, the motivation for agent-based approaches in economic modeling will be outlined (Sect. 3.1). The ACE methodology will be described in more detail in Sect. 3.2. Section 3.3 delivers a survey and critical discussion of the most relevant work in agent-based electricity market modeling, and Sect. 3.4 summarizes the findings from the discussion of the ACE methodology.

3.1 Motivation for Agent-Based Methods in Economics

"[S]tudies of economic systems [– such as electricity markets –] must consider how to handle difficult real-world aspects such as asymmetric information, imperfect competition, strategic interaction, collective learning, and the possibility of multiple equilibria". (Tesfatsion, 2006). Another often neglected factor of electricity markets is that trading is repeated daily. Rothkopf (1999) calls attention to the problem that analyses of electricity markets based on one-shot models of auctions can be far off the mark. In daily repeated auctions bidders will react to what competitors do; the potential for tacit collusion might increase as compared to one-shot auctions.

Many of these factors can not – or only with difficulties – be accounted for with traditional economic modeling techniques. Analytical approaches usually have to put strong and constraining assumptions on the agents that make up the economic system under study, in order to set up elegant formal models.

When the concept of complexity came up, the focus in economic analysis shifted from rational behavior and equilibrium towards heterogeneity and adaptivity (a famous early example being the simulations of Axelrod, 1997). At the same time, the tremendous availability of computational resources made it possible to set up large-scale and detailed computational models that allow a high degree of design flexibility. Populations of heterogeneous agents, feedback from interaction, and dynamic processes attracted notice in computational economic modeling. The first models of this kind have been formulated in the field of evolutionary economics, a research area that deals with phenomena of qualitative change and development (Pyka & Fagiolo, 2005). AB models offered the possibility of not only describing relationships in complex systems, but *growing* them in artificial environments (Epstein & Axtell, 1996). In comparison to traditional scientific methodologies, AB simulation can be seen as a third way between fully flexible linguistic models and more transparent and precise but highly simplified analytical modeling (Richiardi, 2004); the resulting models are dynamic and *executable*, so that its evolving behavior can be observed step by step (Holland & Miller, 1991).

While the foregoing discussion suggests that agent-based modeling is in sharp contrast with analytical modeling, Gulyás (2002) argues that an agent-based implementation is "rather a matter of degree than a binary choice". He presents four reimplementations of a simple model of self-organization. Moving from a top-down approach where firms are represented on an aggregate level (density of firms at one location, market share distribution) through an entity-level representation of the firms towards the introduction of autonomous agents as firm representatives, he can show that even boundedly rational agents can reproduce the basic results gained from the original top-down formulation. However, with each step towards an agent-based implementation, the modeler gains flexibility, as more aspects, like, e.g. heterogeneity of agents, can be accounted for.

Following from the above discussion, an ACE model is a computer-implemented simulation model of a group of agents interacting in an economic situation. It serves for explaining, understanding, and analyzing socio-economic phenomena. According to Tesfatsion (2006), current ACE research can be divided into four strands: The

3.1 Motivation for Agent-Based Methods in Economics   23

empirical, or descriptive strand seeks to understand why and how global regularities on the macro level result from the interaction of economic agents on the micro level. Normative ACE research uses AB models as fully controllable virtual laboratories for testing economic design alternatives in order to determine the policies, institutions, or processes that perform best in an environment of selfish agents. This approach follows the postulation formulated by Roth (2002) that markets should be designed by using engineering tools, such as experimentation and computation. A third strand is theory generation, i.e. the structured analysis of dynamical behaviors of economic systems under alternative initial conditions, in order to find necessary conditions for global regularities to evolve. Finally, ACE researchers continually seek to improve the methodology itself and develop tools that facilitate setting up computational AB models. Most electricity-related research is centered around the second strand, i.e. normative research with the aim of advising good market designs that leave little opportunity for players to exercise market power.

Heterogeneity

One essential element of agent-based modeling is *heterogeneity*. AB modelers are not restricted to equally-sized or symmetric firms, or to other constraints that arise from the limits of analytical modeling. Instead, every agent making up the modeled economy can be designed independently. The economy then evolves as a result of the interplay of these heterogeneous agents, i.e. from the bottom-up.

Heterogeneity can be observed on different levels. Phan (2004) denotes the diversity that agents have by themselves, without any interaction, as *idiosyncratic heterogeneity*. Agents in AB simulations may differ from one another in all sorts of attributes, depending on the research questions at hand, e.g. individual preferences or utility functions, knowledge, attitude towards risk, endowment or wealth, behavioral characteristics or competencies. In contrast, even when agents share the same attribute values at the beginning of a simulation run, their learning capacities together with the specific structure of interaction that they are placed in generally drive the agents towards heterogeneous trajectories, a situation that the same author calls *interactive heterogeneity*.

Heterogeneity is also an important characteristic of real-world electricity markets. Generator agents differ in size and spatial position, they own and operate different generating assets (e.g. fossil fuel fired, nuclear or renewable power plants) with different marginal costs and technical attributes, or they have different strategic characteristics (e.g. vertically integrated or not). During the process of trading, they further differentiate from each other. Agents are engaged in strategic interaction and try to maximize their own profits. They receive information from individual trading success and from market results, and no agent has complete information about the global state of the system. So, each market participant accumulates some knowledge of which strategies might be more or less successful when competing with the rest of the population. On this basis, each agent decides how best to act subsequently (Axtell, 2005). By this means, aggregate system behavior, such as collusion, might be observed. Agent-based simulation is a natural and intuitive way of representing this agent heterogeneity, as it offers much more flexibility in defining each individual market participant than do analytical modeling methods.

Rationality and Adaptation

According to Axtell (2000), one main motivation for agent-based models is the dissatisfaction with rational agents. Thus, he argues, all agent-based models involve some form of boundedly rational agents. In fact, many economists argue that people, although they might try to be rational, can rarely meet the requirement of information or foresight that rational models impose. Already in the 1950s, Simon (1955) proposed to replace the global rationality of *economic man* by some concept of rationality that is compatible with the access to information and computational capacities that humans or firms can actually have. While many researchers agree with this claim, it is less obvious *how* economic agents should realistically be modeled, because – as Holland and Miller (1991) demonstratively state – "[u]sually there is only one way to be fully rational, but there are many ways to be less rational". In his survey, Conlisk (1996) names some approaches of modeling bounded rationality that can be found in the economic literature; as examples, boundedly rational agents may perform limited searches (*satisficing*), apply boundedly rational choice-heuristics (i.e. rules of thumb such as for example *elimination by aspects*), or gradually ameliorate their activities through *learning*, *dynamic adaptation*, or *evolution*.

In classical models, learning and adaptation does not take place because the agents already know everything they need to from the beginning. However, this assumption is not always realistic, and sometimes even not possible. Arthur (1994) presents a problem for which no deductively rational solution can be derived, so that agents can only rely on inductive reasoning. In his very simple model he demonstrates that individual expectations that are boundedly rational can self-organize to produce "collectively *rational*" behavior.[1] He models agents that have to decide whether or not they want to go to a bar in which space is limited and an evening is only enjoyable if it is not too crowded (in his experiments: if no more than 60 persons are in the bar). There is no way to tell the numbers coming in advance; the only information available is the numbers who came in the past weeks. If a person expects fewer than 60 persons to show up, she will go, and she stays at home if she expects more than 60 persons to come. Agents formulate beliefs, or predictors for the number of people going to the bar next week. Well performing predictors

---

[1] An extreme case in this respect has been constructed by Gode and Sunder (1993). In their computerized experiments, *zero-intelligence* traders submit random bids and offers into a double-auction market. The allocative efficiency of the market is observed to be very high (nearly 100%) if a budget constraint is introduced that prevents market participants to sell below their costs or to buy above their values. The authors conclude that market institutions – here, the structure of the double-auction market – can generate aggregate rationality not only from individual rationality but also from individual irrationality.

are used again, and poorly performing predictors are replaced. The surprising result found in computerized experiments of this model is that the mean bar attendance converges to 60, so agents adapt in a desirable way to the aggregate environment they jointly create. This phenomenon might also be applied to other economic situations, such as oligopoly pricing. The conclusion that Batten (2000) draws from presented results is that "[l]earning and adaptation should not be addenda to the central theory of economics. They should be at its core, especially in problems of high complexity".

Many real-world economic situations are complex, and participants have natural limits to perception and information collection, memory, and computational capacity. Adaptation and learning is an obvious representation of human behavior in such situations, and it is one central feature of agent-based models. Holland and Miller (1991) define an agent as *adaptive*, if its actions in its environment can be assigned a value (e.g. performance, utility, payoff, fitness) and if the agent behaves so as to increase this value over time. Adaptive agents are usually not initially endowed with an understanding of the underlying structure of the environment in which they operate. Instead, they must develop a representation of its underlying structure along the course of interaction with the environment and with other agents (Windrum et al., 2007). By means of the information gathered from their environment, agents learn *individually*; simultaneously, through interaction with the environment, each agent contributes also to the adaptation of the whole system, i.e. to *collective learning*.

The question how adaptive behavior can be represented in agent-based models is discussed in more detail in Chap. 4, where especially the concept of reinforcement learning is considered.

3.2 Building Agent-Based Simulation Models

In the following, the basic concepts of agent-based computational models will be described in more detail, and some suggestions for the design process of building such models will be given (Sect. 3.2.1). Validation and calibration of AB models is a specifically demanding task within this process; it will therefore be discussed explicitly in Sect. 3.2.2. Section 3.2.3 finally presents some software toolkits that facilitate the agent-based simulation process.

3.2.1 Methodology and Main Concepts

The ACE modeling procedure can be described as follows (Tesfatsion, 2002): After having (1) defined the research questions to resolve, the ACE modeler (2) constructs an economy comprising an initial population of agents and subsequently (3) specifies the initial state of the economy by defining the initial attributes of the agents (e.g. type characteristics, learning behavior, knowledge about itself and other agents); the modeler then (4) lets the economy evolve over time without further intervention – all events that subsequently occur must arise from the historical time-line of agent-agent interactions, without extraneous coordination; this procedure is followed by (5) a careful analysis of simulation results and an evaluation of the regularities observed in the data.

Actors in AB models are represented by computational agents. As the term "agent" is not used in a consistent manner by different disciplines, this section starts by briefly introducing the notion of agents in the context of Agent-Based Computational Economics.

Software Agents

The concept of (computational or software) "agents" stems from the fields of Distributed Artificial Intelligence (DAI) and Multi-Agent Systems (MAS). Common definitions of the term characterize them as *autonomous, reactive, goal-oriented*, or *socially able*, just to cite a few (for a discussion of the term, see e.g. Franklin & Graesser, 1997). However, as Drogoul et al. (2003) correctly annotate, these features do not all translate into computational properties in agent-based simulations. Most AB models do not require agents to exhibit all the characteristics of the software agents from the DAI or MAS world; instead, the most important features of agents in AB models is that they are *goal-oriented* in that they seek to maximize an assigned value like payoff, fitness, or utility, and they are *adaptive*, meaning that they learn which actions to take in order to increase this value over time, and so reach the goal.

In AB electricity simulations, the most common agents that make up the population are generators, load serving entities, and a market/system operator. Depending on the research questions, the simulation can also contain regulator agents, a transmission system representation, retail customers, or others. Agents can also be composed of other agents, thus permitting hierarchical constructions, like utilities or trusts.

Design Process

In order to enrich the understanding of fundamental processes and dynamics, one important principle of economic modeling also applies to ACE modeling, i.e. to keep models as simple as possible. In the context of agent-based modeling, Axelrod (2006) puts it as follows: "*While the topic being investigated may be complicated, the assumptions underlying the agent-based model should be simple. The complexity of agent-based modeling should be in the simulated results, not in the assumptions of the model*". If the goal of ACE research is to deepen our understanding of some fundamental process, then simplicity of the assumptions is much more important than realistic representation of all the details of a particular setting. Simplicity is

3.2 Building Agent-Based Simulation Models

especially important in agent-based simulations when unintuitive results occur. In this case, it is helpful to understand everything that goes into the model in order to be sure that the results can be explained by the model's characteristics and do not stem from implementation errors.

There are only few guidelines proposed for the design and implementation process of agent-based simulation models. As a start, basic principles of simulation procedures used in other disciplines also apply to agent-based simulation. Law (2007) for example provide helpful guidance for building simulation models and analyzing output data; Gilbert and Troitzsch (2005) describe simulation methodologies in the social sciences, including some hints for agent-based modeling.

Drogoul et al. (2003) formulate one proposition for the design process of agent-based computational models (see Fig. 3.1 for an overview). They describe the way how a *target system* which characterizes the phenomenon to predict or the theory that needs explanation can be translated into a *computational system*. Their methodology also suggests an iterative procedure of simulation runs with different scenarios, analysis and interpretation on the one hand, and model refinement on the other hand. The simulation process is described through different "roles" that are involved in the design process:

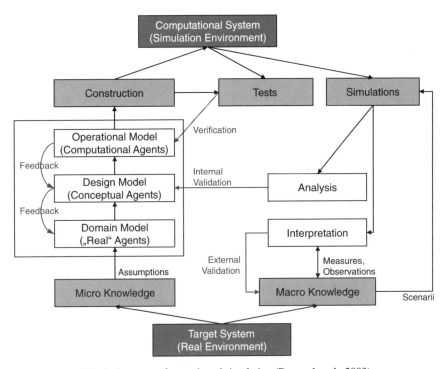

Fig. 3.1 Methodological process of agent-based simulation (Drogoul et al., 2003)

- The intention of the simulation is defined by the *thematician*, who formulates the research questions that should be answered with the model and characterizes theories and assumptions that describe the considered domain. The thematician translates *micro knowledge*, i.e. knowledge about individuals and their behavior, into the *domain model* which contains the agents. The *macro knowledge*, i.e. "global" knowledge about the target system, is used to provide scenarios and measures for the simulation.
- The next modeling step is effected by the *modeler*, who is responsible for formulating the *design model* on the basis of the *domain model*. A *design model* is more formalized, so that it can be translated into an executable computer program. It can be compared to a Unified Modeling Language (UML) diagram in object-oriented programming.
- The *computer scientist* is involved in translating the *design model* into a computer program.

The whole process of agent-based modeling contains several iterations and feedback loops between the target system and its computational implementation. Verification and validation steps make sure that the model is correctly implemented and that it represents the real world in an appropriate way. Simulation runs with different scenarios then help to gain the desired insights of the simulated system and answer the research questions, but also to further refine the model. This whole process is depicted in Fig. 3.1.

While the presented model building process offers some guidance, it does not give any advice as to how the developed simulation model can be categorized and described in a publication, and how it should be calibrated and validated. In order to make AB simulations better understandable and publishable, Richiardi, Leombruni, Saam, and Sonnessa (2006) have made some practical suggestions towards a standardized methodological protocol of AB simulations; they propose to:

- Include references to the theoretical background of the economic phenomenon that is investigated, including simulation and non-simulation literature
- State the main features of the simulation model (treatment of time: discrete or continuous, treatment of fate: stochastic or deterministic, coordination structure: centralized or decentralized, among others) clearly and immediately in order to facilitate understanding and model comparison
- Follow well-defined processes for data analysis, including accepted calibration, validation and sensitivity analysis techniques
- Use standard modeling languages such as the Unified Modeling Language (UML) for describing the static and dynamic properties of the model and use AB modeling toolkits in order to make models more easily replicable

Some examples for toolkits that facilitate agent-based modeling are presented later in this chapter. The procedure of model calibration and validation will be discussed in the following section.

3.2.2 Validity of Agent-Based Simulation Models

The difficulty of validating ACE model outcomes against empirical data is one of the weaknesses of the ACE methodology (Tesfatsion, 2006).[2] This point can clearly be observed when surveying the agent-based electricity modeling literature (see also discussion at the end of this chapter, where some suggestions for further research into this problem are given). Among others, Windrum et al. (2007) also find fault with the lack of comparability between AB models that have been developed. They observe that many AB researchers do not undertake an in-depth analysis to compare and evaluate their relative explanatory performance. Rather, models are viewed in isolation from one another, and validation involves, at the maximum, examining the extent to which the output traces generated by a particular model approximates one or more "stylized facts" drawn from empirical research.

Building sound, valid, and credible simulation models entails several procedures. First, one has to ensure that the computational model is correctly implemented and working as intended. This procedure is referred to as *verification* and mainly consists of debugging the implemented computer program. *Calibration* is referred to as the process of choosing parameter values that maximize the correspondence of the model's output with measured data from the real-world system. Finally, in the *validation* step, the modeler has to check whether the simulation model accurately represents the actual system being analyzed, from the perspective of the research objectives that the model is applied to. More precisely, *model validity* can be subdivided into different levels, all of which have to be considered in the process of agent-based simulation (Richiardi et al., 2006):

- *Theory validity* – validity of the theory relative to the real-world system
- *Model validity* – validity of the model relative to the theory
- *Program validity* – validity of the simulation program relative to the model
- *Operational validity* – validity of the theoretical concept relative to its indicators
- *Empirical validity* – validity of the empirically occurring true value relative to its indicators

Another important procedure in assuring the model's quality is to find out how sensitive the model is to slight changes in parameter values and initial conditions; this step is referred to as *sensitivity analysis*. The procedures presented here do not necessarily have to be executed consecutively; loops and jumps between the steps may become necessary along the development of different versions of the model.

Many AB models lack some of the steps enumerated above, which may cause a skepticism towards the methodology. There is a considerable body of literature that suggests guidelines and techniques for simulation verification and validation (e.g.

---

[2] It should be noticed that some researchers argue that AB models are only suitable for qualitative analysis. This would entail that AB models can solely test theories in the form of causal relationships, and calibration would be less an issue. Validation could not be grounded on a comparison with empirical data in this case (see Pyka & Fagiolo, 2005; or Windrum et al., 2007 for this discussion). In contrast, most of the latest papers on AB model validation consider empirical validation.

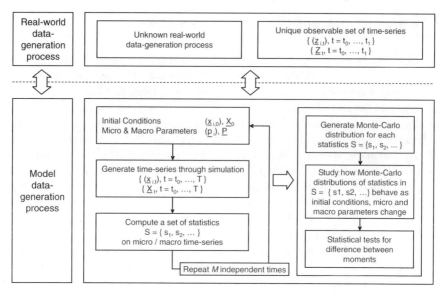

Fig. 3.2 Procedure for studying the output of agent-based simulations (Windrum et al., 2007)

Gilbert & Troitzsch, 2005; Law, 2007; Sargent, 2005). Very recently, the need for reliable validation techniques has obviously been recognized. AB researchers have analyzed and suggested procedures and guidelines for calibrating and validating agent-based simulation models, e.g. Windrum et al. (2007), Marks (2007), Richiardi et al. (2006), Midgley, Marks, and Kunchamwar (2007), and a whole journal special issue is devoted to this topic (Fagiolo et al., 2007).

Windrum et al. (2007) review some empirical validation techniques that could be used to ensure validity of AB models. They first examine how the output of an AB simulation can be analyzed (see Fig. 3.2). Here, $\underline{z}_{i,t}$ corresponds to a set of empirically observed data at time t on the agent (micro) level, and $\underline{Z}_t$ on the aggregate (macro) level. Similarly, $\underline{x}_{i,t}$ denotes the agent-level and $\underline{X}_t$ the aggregate simulation outcome. The model is run until it reaches some form of stable behavior for a particular set of initial micro and macro parameter values ($\underline{p}_i$, $\underline{P}$, $\underline{x}_{i,0}$, and $\underline{X}_0$). The model outcome is characterized by a set of statistics $S = \{s_1, s_2, ...\}$ that are computed from data generated by the model data generation process. As most processes in AB simulations are stochastic in nature, several simulation runs with varying random number seeds are necessary. These represent several different possible outcomes, e.g. possible ways in which agents respond to the actions of other agents. For any given run $m = 1, 2, ..., M$, the simulation will output a value for $s_{j,m}$, which will be different across runs. Producing M independent runs leads to a distribution for s_j with M observations. From this distribution, the mean values $E(s_j)$, or its variance $V(s_j)$ can be calculated. These depend, of course, on the initial choices that were made for $\underline{p}_i$, $\underline{P}$, $\underline{x}_{i,0}$, and $\underline{X}_0$. By exploring a sufficiently large number of points in the space of initial conditions and parameter values, and by computing $E(s_j)$, $V(s_j)$, and other statistics at each point, one can gain an understanding of the behavior of the model data generation process.

In order to ameliorate the problem of empirical model validation, LeBaron (2006) makes three suggestions (here, the author's focus is on financial markets): The first is to attempt to construct an AB model such that it replicates empirical features which are not well replicated by standard models. The second one is to put as many parameters as possible under evolutionary control in order to find optimal values for crucial parameters. The third suggestion is to use insights gained from experimental economics in order to build more realistic learning dynamics.

Moss and Edmonds (2005) propose to *cross-validate* AB models. They argue that the micro level of an AB model is best investigated qualitatively while the macro level should be investigated using quantitative methods. When the agents' behavior and interaction on the micro level is able to generate the macro level phenomena sharing the statistical characteristics of data from the real-world system, then the model is cross-validated: the micro level behavior has been validated qualitatively by domain experts, and the macro level data have been validated by comparing statistical properties of numerical outputs from the model with statistics of the real-world system.

Werker and Brenner (2004) propose an advanced methodology for calibrating and validating AB simulation models based on *Critical Realism*. It proceeds in three steps which involve (1) setting possible model specifications (parameters, interactions), where the assumptions on which the model is built should be induced from empirical data whenever this is possible; a set of several plausible models with certain parameter ranges results from this first step; (2) running each model specification many times and comparing the tested model specifications to empirical observations, rejecting those that are not confirmed by the empirical data; (3) identifying the underlying mechanisms driving the part of the world that should be described and explained, using abduction.

From the presentation provided here, it can be seen that in fact some recent developments in the AB method have culminated into guidelines and procedures for improving the reliability and validity of agent-based simulation models. This is a necessary step for reducing skepticism and ensuring the quality of the methodology.

3.2.3 Software Toolkits for Agent-Based Simulation

Software development is a tedious and difficult task in the ACE research procedure. When a considerable number of researchers rely on agent-based models, much can be gained from having standard procedures and basic agent-based simulation functionalities available. As a consequence, a wide variety of agent-based simulation toolkits or platforms have been developed in an attempt to facilitate model development and implementation. These toolkits provide standard concepts for designing and describing agent-based simulations, along with the source code or libraries that deliver standard functionalities for implementing the framework. Libraries usually include objects for creating, running, and displaying simulation runs, and for collecting generated data along the run. The discussion of which platforms are best suited for Agent-Based Computational Economics models is briefly summarized in the following.

Gilbert and Bankes (2002) give a short overview of agent-based modeling platforms and tools which includes Repast,[3] Swarm,[4] Ascape,[5] and some "visual programming" tools that only allow the definition of very simple models. Tobias and Hofmann (2004) compare the usefulness of four different toolkits for agent-based social simulation, i.e. Repast, Swarm, Quicksilver[6] and VSEit[7] (all toolkits are based on the Java programming language). A recent comparison of agent-based modeling platforms, conducted by Railsback, Lytinen, and Jackson (2006), describes the main features and characteristics of NetLogo,[8] MASON,[9] Repast, and the Java and Objective-C versions of Swarm. Another survey of a large number of AB software toolkits is provided by van Dinther (2007), with some tools, namely Ascape, JADE,[10] MadKit,[11] Repast and Swarm being analyzed in more detail. Based on the comparison of the toolkits and on criteria derived from the detailed study, the author develops his own simulation tool, the *Agent-based MArket Simulation Environment* AMASE. Finally, Prof. Tesfatsion maintains a web page listing *General Software and Toolkits for Agent-Based Computational Economics and Complex Adaptive Systems.*[12]

Tobias and Hofmann (2004) ranked all reviewed toolkits according to three classes of criteria that they judge relevant for social scientist engaged in agent-based modeling. These are (1) general criteria such as licensing, documentation and support, and the tool's future viability; (2) modeling and experimentation criteria that help to answer the question how good a tool supports the development of basic simulation features and ensures the program's efficiency so that the researcher can concentrate on theory and content of the model; and finally (3) the modeling options that the toolkit supports, such as for example agent communication, possible agent complexity, management of spatial arrangements. The authors conclude from their comparison that Repast outperforms the other three considered tools in almost all analyzed criteria.

Railsback et al. (2006) focus primarily on the *ease of use* of the different toolkits. They ask how difficult it is – for an experienced and for an unexperienced programmer – to implement agent-based simulation models and conduct experiments on them. Another quality criterion for the authors is the execution speed of simulations. Their conclusion is less definite. NetLogo is found to be easy to use and well documented. They consider it well suited for less complex models of short-term, local interaction of agents and a grid environment. Objective-C Swarm, one

---

[3] http://repast.sourceforge.net/; see also (North et al., 2006).

[4] http://www.swarm.org/

[5] http://ascape.sourceforge.net/

[6] http://quicksilver.tigris.org/

[7] http://www.vseit.de/

[8] http://ccl.northwestern.edu/netlogo/

[9] http://cs.gmu.edu/ eclab/projects/mason/

[10] http://jade.tilab.com/

[11] http://sourceforge.net/projects/madkit/

[12] http://www.econ.iastate.edu/tesfatsi/acecode.htm

of the first agent-based modeling platforms, is considered stable, relatively small and well-organized while providing a fairly complete set of tools. However, the authors criticize its low availability of documentation and tutorial materials. Given that alternative good Java toolkits are available, the Java version of Swarm offers no real advantages in the opinion of the authors, and is not advised. Repast was identified to be the most complete Java toolkit with good execution speed and many helpful functions, although accessibility and documentation of the code distributions were found to be disappointing. MASON has shown to be the fastest platform and is judged well appropriate for experienced programmers working on models that are computationally intensive.

This short overview shows that there are a number of toolkits available that ease the process of model implementation. They increase reliability and efficiency of simulation, as complex procedures have been developed by programming experts. This also helps to harmonize ACE modeling procedures and to focus on the content and those parts of the model that really matter. Toolkits can facilitate the replicability of AB simulations and are clearly favorable as opposed to building models from scratch. However, none of the tools seems to be suitable for all kinds of simulations, but adequacy depends on the specific modeling purpose.

3.3 Related Work: ACE Electricity Market Models

A growing number of researchers have developed agent-based computational models for simulating electricity markets. The modeling approaches and findings described in the most relevant studies are reviewed in the following. This review concentrates on agent-based simulation models for the analysis of market structures and market design in wholesale electricity trading. Aside from this, some researchers have applied agent-based methods for examining electricity consumer behavior at the retail level, e.g. Hämäläinen, Mäntysaari, Ruusunen, and Pineau (2000), Roop and Fathelrahman (2003), Roop, Fathelrahman, and Widergren (2005), Yu et al. (2004), or Müller, Sensfuß, and Wietschel (2007); others provide agent-based decision support tools for power market participants, e.g. Praça, Ramos, Vale, and Cordeiro (2004), Bernal-Agustín, Contreras, Martín-Flores, and Conejo (2007), or Harp, Brignone, Wollenberg, and Samad (2000). These will not be discussed here.

A more detailed literature survey of agent-based wholesale electricity market models is provided by Weidlich and Veit (2008c).

3.3.1 Simulations Applying Reinforcement Learning

A considerable proportion of agent-based electricity simulations described in the literature use some form of reinforcement learning for representing agent behavior. The most popular among these are the Erev and Roth reinforcement learning algorithm and Q-learning.[13]

---

[13] These learning algorithms with their relevant parameters and possible designs are introduced in detail in Sects. 4.1.1 and 4.1.2.

Developing the Erev and Roth Algorithm (Nicolaisen et al.)

Petrov and Sheblé (2001), Nicolaisen, Petrov, and Tesfatsion (2001), and Koesrindartoto (2002) experiment with simple electricity market models applying Erev and Roth reinforcement learning. The authors discovered some problematic features of the original Erev and Roth (1998) algorithm formulation, i.e. that no propensity update occurs when profits are zero or for some values of action domain size and experimentation parameter. These considerations lead to the formulation of the *Modified Roth-Erev* algorithm (MRE), which is used in several AB electricity papers. Nicolaisen et al. (2001) apply the MRE for simulating a double-auction electricity market with discriminatory pricing. The aim of the study is to analyze market power and efficiency as a function of relative concentration and capacity of the market. Simulation results do not support the hypotheses that the market power of sellers increases when relative capacity increases or when relative concentration decreases. However, results support the hypothesis that market efficiency (total profits in the simulated auction in relation the competitive equilibrium outcome) is high. The authors compare their results to the market efficiency observed simulations using Genetic Algorithms (Nicolaisen et al., 2000, see also Sect. 3.3.2). They come to the conclusion that individual MRE learning leads to higher market efficiency than GA social mimicry learning, because each agent learns to ameliorate his own bidding strategies given his individual cost structure, instead of mimicking strategies of structurally distinct traders.

Wholesale Market Reliability Testing with AMES (Tesfatsion et al.)

Koesrindartoto and Tesfatsion (2004), Koesrindartoto, Sun, and Tesfatsion (2005), and Sun and Tesfatsion (2007) describe an electricity market model that encompasses the core features of the *Wholesale Power Market Platform*, a market design proposed by the U.S. *Federal Energy Regulatory Commission*. The *AMES* model (*Agent-based Modeling of Electricity Systems*, Sun & Tesfatsion, 2007) comprises a two-settlement system consisting of a day-ahead market and a real-time market which are both cleared by means of locational marginal pricing. The market clearing is managed by an independent system operator (ISO) agent who operates an AC transmission grid.[14] The authors report on initial simulation results that are run with a 5-node transmission grid test case and with only one day-ahead market (the real-time market is inactive). The demand side is simplified to a fixed and price insensitive daily load profile submitted to the ISO. Generator agents learn to optimize a supply function through strategically setting "reported marginal costs" (bid prices) at the minimum and maximum possible capacity levels. For this purpose, they apply the MRE algorithm with Softmax[15] action selection.

---

[14] The representation of the transmission grid in this model is approximated by a bid-based DC optimal power flow (OPF) problem which is described in more detail in Sun and Tesfatsion (2007).

[15] See Sect. 4.1.1 for a description of Softmax action selection.

3.3 Related Work: ACE Electricity Market Models   35

The simulation results show that all five generator agents learn to successfully submit bids above their true marginal cost. This leads to total variable costs of operation that are about three times higher then they are in the case where generators report their true marginal costs. The authors conclude that the *Wholesale Power Market Platform* design features do not prevent the considerable exercise of market power by generators. Further extensions of the *AMES* model are envisaged.

Vertical Integration in the Energy Sector (Rupérez Micola et al.)

Rupérez Micola, Banal Estañol, and Bunn (2006) present a model that consists of three sequential oligopolistic energy markets representing a wholesale gas market, a wholesale electricity market and a retail electricity market. They analyze the effect of reward interdependence in vertically integrated energy firms. Trading in all three markets is modeled as a uniform-price auction with fixed inelastic demand, to which seller agents submit their bids. Agents always bid their full capacity and strategically set bid prices; possible actions (prices) range from 0 to an upper price ψ in the retail market (which is cleared first), from 0 to the resulting retail price in the wholesale electricity market, and from 0 to the resulting wholesale electricity price in the gas market. Vertically integrated firms are modeled as two agents that each trade in one market and have interdependent rewards: The reinforcement that an agent perceives consists partly on his trading result in one market, and partly on the profits earned in the interlinked market, the fraction being denoted as the *reward interdependence parameter* $\alpha = \{0, 0.01, 0.02, ..., 0.5\}$. Agents apply a slightly modified Erev and Roth learning algorithm.

Results for simulations with two gas shippers, three wholesale electricity traders, and four retail electricity traders show that the presence of vertically integrated firms generally raises prices in at least two of the three markets considered. The authors show that the vertically integrated firm can increase its overall profits. However, the disintegrated firms in the gas and wholesale electricity market can also increase their profits as a result of higher prices in the scenarios with high reward interdependence.

In contrast to the authors' presumption that agents can achieve higher profits through coordinating overall profits in both markets ("vertically integrated firms give up profits downstream in order to increase the scope for upstream profits"), the agents' learning model does not allow developing strategies across several markets. The reason for increasing prices might lie in the absolute order of reinforcements: profits are generally higher in the gas market than in the electricity market. As a consequence, vertically integrated gas shipper agents whose reinforcements only contain a part of the profits earned in the gas market (the other part coming from the electricity market), are inevitably less satisfied with their trading results than their (otherwise identical) rivals. Thus, they search for other ways of attaining higher profits, leading to higher prices in the gas market. When prices in the gas market rise, margins in the wholesale electricity market shrink, so agents may seek to compensate this loss through higher bid prices. This highlights the importance to explain observed macro behavior only with causes that lie in the agents' learning tasks.

Further Approaches Applying Erev and Roth Reinforcement Learning

Bin, Maosong, and Xianya (2004) report on simulation results comparing three different pricing methods in electricity auctions. Agents in their model submit bids for their whole installed capacity; they learn to bid mark-ups on top of their marginal costs. The learning algorithm applied is similar to the Erev and Roth reinforcement learning algorithm with proportional action selection. The authors compare uniform pricing, pay-as-bid pricing and a mechanism called *Electricity Value Equivalent* pricing. Simulations are carried out for two cases, i.e. one in which each generator agent owns only one power plant and the second in which the same capacity belongs to less generators, each owning more than one power plant.[16] Both the model and the result presentation are rather brief and in parts confusing; the conclusion of the simulation results is that the third pricing mechanism leaves less room for generators to exercise market power than the other two considered methods.

Cincotti, Guerci, and Raberto (2005) model a day-ahead electricity market which takes the form of a clearing-house double-auction with uniform pricing. While the authors have used the original formulation of the Erev and Roth reinforcement learning algorithm in earlier work (Cincotti & Guerci, 2005), they now propose a new algorithm, which they say is inspired by Erev's and Roth's original work. The learning formulation they propose not only evaluates the profits gained from chosen actions, but also evaluates potential profits that would have been realized had other actions been chosen.[17] Agents learn to bid price quantity pairs that maximize their profits; bid prices can range from the generator's marginal cost to the maximum admissible price, and bid quantities range from 0 to the generator's maximum installed capacity, both in steps of 1 EUR, or 1 MW respectively. A strategy is deleted once it does not perform better than the currently played strategy. This leads to a rapid and unjustified constriction of the action space and may hinder agents from finding the best strategies. The authors conduct experiments for different cases of supply side competition. Among the findings is that prices quickly converge to the competitive equilibrium value (i.e. marginal cost) in most cases. The authors find that the equilibrium outcome is reached faster when the number of competing generators is small. This finding is rather unintuitive; it may be an indicator for the model not being a realistic representation of the market under study.

Weidlich and Veit (2006) simulate two markets that are cleared sequentially – a day-ahead electricity market and a market for balancing power. Agents place bids in both markets and evaluate their individual success in one market by integrating the opportunity cost of profits that could have been obtained in the other market. They learn from trading results using a modified Erev and Roth reinforcement learning algorithm with proportional action selection. The day-ahead market is modeled as a call market with a fixed and price-insensitive demand side. At the balancing power market both pay-as-bid and uniform price settlement have been

---

[16] It is not specified in the paper how agents strategically bid a portfolio of power plants in the second case.

[17] Potential profits are calculated under the assumption that an agent would have sold the total bid volume if his bid price had been less than or equal to last round's market clearing price.

simulated. The authors show that the order of market execution plays a significant role in resulting market prices: if the day-ahead market is cleared first, agents still have more available capacity to offer in the auction. Hence, competition is increased and prices are lower as compared to the case in which some generators have already sold (parts of) their capacity in the balancing power market. As potential profits that could have been earned on the day-ahead market play a more important role for setting bid prices in the balancing power market than vice versa, low prices in the day-ahead market also lead to lower prices in the balancing power market. When the balancing power market is cleared first, prices are higher in both markets. The authors also observe that agents tend to bid lower prices in the case of uniform pricing; however, resulting (average) prices are lower in the pay-as-bid case.

Veit et al. (2006) study the dynamics in two-settlement electricity markets. In these markets, energy producers sign strategic contracts in the forward market, and engage in oligopolistic competition in the spot market. While electricity is traded in a few trading zones of the network in the forward market, the spot market considers a more detailed underlying transmission network in form of a lossless DC power flow optimization problem. In the spot market, electricity is paid at nodal prices; forward contracts are settled at spot zonal settlement prices. Agents learn to set profit maximizing bids on the spot and forward market separately. They apply a modified Erev and Roth reinforcement learning algorithm with proportional action selection. Bids are evaluated individually for each power plant and for each possible spot market state. In contrast, propensities for forward bids are updated on the basis of the generators' total profit, i.e. learning is not differentiated for single plants. The agents' learning task is to set profit maximizing bid quantities on both markets (Cournot game). The load is modeled as a linear demand function. Simulation results from this model demonstrate that the introduction of a forward market leads to a more competitive bidding behavior by suppliers in the spot market, and thus to lower spot electricity prices.

Q-Learning Model of Electricity Markets with Network Constraints (Krause et al.)

Krause et al. (2005) compare Nash equilibrium analysis and AB modeling with Q-learning for the case of a power pool with transmission constraints (the network representation is depicted in Fig. 3.3). The model contains three agents who strategically set mark-ups to their electricity selling bids. The agents apply an update rule similar to the one used in Q-learning, but without differentiation between the states that agents are in at each iteration: $Q(a_t) \leftarrow Q(a_t) + \alpha(r_{t+1} - Q(a_t))$. Despite the simplification, the authors achieve satisfactory simulation results. They compute Nash equilibria for simplified model settings. In those cases where there exists one Nash equilibrium, the agents' actions quickly converge to this equilibrium, whereas in the case of two Nash equilibria, cyclic behavior is observed.

In a later study, Krause and Andersson (2006) apply the same model for analyzing the social welfare implications of three different congestion management methods, i.e. *locational marginal pricing* (LMP), *market splitting*, and *flow-based*

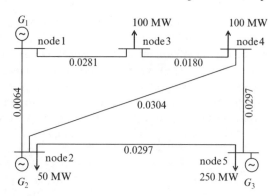

Fig. 3.3 Transmission system in the model described by Krause et al. (2005)

market coupling (FBMC). The case in which all generators bid their true cost functions (referred to as *perfect competition*) is compared to a case in which all three generators apply the above described Q-learning-like algorithm (the *oligopolistic competition* case). Simulation results reveal that LMP results in the highest overall welfare in both the competitive and the oligopolistic case. Market splitting performs second best in both cases and FBMC shows lowest overall welfare. However, in the case with learning agents, generators' surplus is higher while consumer surplus is lower than in the competitive case. This is due to the exercise of market power by the generators.

Further Approaches Using Q-Learning

Similar results as the ones reported precedently have been found by Naghibi-Sistani, Akbarzadeh-Tootoonchi, Javidi-Dashte Bayaz, and Rajabi-Mashhadi (2006). In their model which differentiates between the two environmental states of "low production cost" and "high production cost", agents strategically set the slope of their linear bid supply function. Possible bidding strategies are classified as high, middle and low (price/slope). Two bidders compete to satisfy demand in a power pool without transmission constraints. They apply the Q-learning algorithm with Softmax action selection. The temperature parameter is adjusted during the simulation according to a formula in which the temperature is high when the Q-values have not converged to a final value, and low when changes in Q-values become small. The motivation behind this formulation was to reduce convergence time and sensitivity to learning parameters. However, the formulation seems rather arbitrary and is even counterproductive, because a new parameter is introduced. Reported results from an exemplary simulation run show that agents quickly learn to play the strategy corresponding to a Nash equilibrium; the probability of these strategies converge to 1 after ∼100 iterations.

Xiong, Okuma, and Fujita (2004) compare market prices and price volatility in uniform price and pay-as-bid electricity auctions with the help of a model with generator agents applying Q-learning. The environmental states are defined as the last round's market prices. Agents learn to set bid prices that maximize their payoffs, and bid volumes are always equal to the net capacity of a generator. The reinforcement for each hour h of the day comprises the profit r of that hour, and also the ratio of actual and target utilization rate (p is a parameter defining how strictly an agent tries to satisfy the target utilization rate): $r_h(s,a) = r(h) \times \left(\frac{\text{actual utilization}(h)}{\text{target utilization}(h)}\right)^p$. Action selection is effected with the ε-greedy strategy. Simulation runs are conducted with both price-inelastic demand and a responsive demand-side; both demand scenarios are run with uniform price and pay-as-bid clearing. The authors compare market prices[18] and bid prices of one agent for the four simulated scenarios. The conclusion from their simulation results is that agents bid at higher prices in the pay-as-bid case, but overall prices are higher with uniform pricing. The introduction of interruptible loads causes prices to drop for both clearing mechanisms; however, the price decrease is stronger in the uniform price case.

Bakirtzis and Tellidou (2006) also compare uniform and pay-as-bid pricing with agents applying Q-learning. Environmental states are defined by the last round's market price (uniform price or the quantity-weighted average price of winning bids, respectively). Agents also bid their full capacity at the price learned through Q-learning; possible bid prices are bounded by the generator's marginal generating cost (floor) and the price cap (ceiling). The action selection rule applied is adapted from the *Simulated Annealing Q-learning* algorithm developed by Guo, Liu, and Malec (2004). Load is assumed to be constant and price-inelastic. Four cases are examined, varying between uniform and pay-as-bid pricing with either one dominant generator with lowest marginal costs whose installed capacity is enough to satisfy total demand alone, or three equally sized generators instead of the dominant firm. Results show that the dominant firm learns to set the bid price that fully maximizes its profits with both pricing rules. The two next cheapest suppliers learn to stay below the bid price of the dominant firm, in order to be dispatched. In the pay-as-bid case, the bid prices of these two generators are closer to the dominant generator's bid than under uniform pricing, but resulting market prices are slightly higher in the latter case. In the more competitive cases, the three generators with the lowest marginal generating costs compete to serve demand, and prices mostly stay below the marginal cost of the next cheapest generator. Here, however, average prices under pay-as-bid are higher than market clearing prices under uniform pricing.

3.3.2 Simulations Applying Evolutionary Concepts

Among evolutionary adaptation models (see Sect. 4.2 for a short introduction), especially genetic algorithms (GA) and learning classifier systems (LCS) have been applied to electricity market simulations. Curzon Price (1997) has demonstrated the

---

[18] It is not specified how the "market price" is calculated in the pay-as-bid case, in which each generator faces a different price. It can be assumed that the authors take the weighted average over all successful bids as the (average) market price.

usefulness of GAs for simple standard games such as Bertrand and Cournot competition, price choice of a monopolist and a chain of monopolists, and also for very simplistic electricity market settings. In the following, some GA and LCS simulation models designed specifically for electricity market research are summarized.

Early GA Approaches (Richter et al.)

Richter and Sheblé (1998), Petrov and Sheblé (2000), Lane, Kroujiline, Petrov, and Sheblé (2000), and Nicolaisen et al. (2000) present simple electricity market models that use genetic algorithms for representing the agents' bidding behavior. The simulated market in the cited papers takes the form of a double-auction where executable supply and demand bids are matched pairwise. Prices are determined as the midpoint between two matched bids or as a *competitive equilibrium* price. The learning task for the agents differs in the cited papers. Richter and Sheblé (1998) define only generation companies as part of the GA population. They have three distinct evolving parts, or genes: one for determining the offer quantity, another for selecting the offer price and a third one for choosing a price forecast method. The latter include strategies like e.g. moving average or linear regression. On the basis of the respective technique coded in their genes, agents determine their forecast price, then choose a bid price between their generating cost and the forecast equilibrium price, and an offer quantity between 0 and their maximum generation capacity. Standard GA methods are used for the evolution of the population.

Petrov and Sheblé (2000) propose a model where only one agent applies a genetic algorithm for evolving its trading strategy. All other electricity sellers (buyers) increase (decrease) their bid price by one increment if the last offer was accepted; in the opposite case, they decrease (increase) their bid price for the next round. The GA agent develops more sophisticated trading strategies based on the last round's bid price and the equilibrium price. The strategy has the form of a decision tree whose functions comprise algebraic and logical operators (e.g. summation, division, greater, "if-then-else"). The authors find that the GA agent consistently surpassed the fixed-rule agents during all separate runs of the simulation.

In Nicolaisen et al. (2000) and Lane et al. (2000) both buyer and seller strategies – here bid and ask prices – evolve with the help of a genetic algorithm. The bit string representing the agent's gene codes a floating point number in the interval $[0, 1)$ which is multiplied by a dollar constant, resulting in the admissible bid (ask) price. The authors are interested in measuring the individual market power of buyers and sellers. When discussing their simulation results (most of which do not confirm their formulated hypotheses and even contradict theoretical economic considerations), the authors admit that their very simple implementation of GA might not realistically represent human behavior in real markets. They propose to apply learning representations which allow agents to learn on the basis of their individual experience in the trading process. In later studies by the authors (Petrov & Sheblé, 2001; Nicolaisen et al., 2001, see Sect. 3.3.1), reinforcement learning is used for representing the agent behavior.

3.3 Related Work: ACE Electricity Market Models

Models of the Australian National Electricity Market (Cau et al.)

Cau and Anderson (2002) develop a wholesale electricity market model similar to the Australian *National Electricity Market*. In Cau (2003), the model is developed further. It covers two bidding structures: stepwise and piecewise linear bidding. In both cases, agents assign bid quantities to a number of given price segments. In the piecewise linear case, bidding schedules are formed by linear interpolation between two bid points, whereas quantities are kept constant between two points in the stepwise case. The agents' task is to find the strategies that maximize their individual payoffs. A strategy is a set of bidding schedules for the possible environmental states. A state is defined by the previous spot market price, past market demand and forecast market demand. Demand is price-inelastic and can be classified as either *high* or *low*, with some uncertainty about the exact level. The fitness of a strategy is measured as the average payoff achieved. Standard evolutionary operations, such as crossover and mutation, are effected on the population. Simulations are run for a duopoly with two equally sized generators with different marginal generation costs. The authors observe that tacitly collusive strategies can be learned by the agents in this co-evolutionary environment, both in the stepwise and piecewise linear bidding structure. High overall demand, high demand uncertainty and low price elasticity facilitates tacit collusion (measured as the joint profit ratio). On the supply side, situations in which tacit collusion is easier to achieve are characterized by symmetry in cost and capacity, and small hedging contract quantity. Besides, a high number of agents makes it more difficult for them to collude in a sustainable way. However, even in cases with many competing generators, tacit collusion can still occur.

LCS Approaches to Electricity Market Modeling (Bagnall and Smith)

Bagnall and Smith (2005) present a simplified AB simulation model of the pre-NETA[19] UK electricity market that applies LCS for developing successful bidding strategies. The model and simulation results have also been described in several other papers, e.g. A. Bagnall and Smith (1999), Bagnall and Smith (2000), A. J. Bagnall (2000b), Bagnall (2000a), and A. Bagnall (2004). The agent population in the presented model consists of 21 generator agents, each owning one generating unit of one of the four types nuclear, coal, gas, or oil/gas turbine. Every generation type has different characteristics of fixed cost, start-up cost, and generation costs. The market environment is characterized by the forecast demand for each half-hour time slot, by capacity premia for each half-hour time slot of the following day, and by transmission constraints. During one daily trading cycle, the system marginal price and pool purchase price[20] are calculated from all bids submitted by the generators (unconstrained schedule). Constraint levels, together with the unconstrained schedule, result in the constrained schedule. The agents' objectives is to avoid losses

---

[19] NETA = New Electricity Trading Arrangements.

[20] The pool purchase price is the sum of the system marginal price and the capacity premium.

and to maximize profits. They apply one learning classifier system for each objective. The learning task is to set the bid price. Each LCS has three components: the *performance component* produces a prediction array consisting of estimates of the expected reward for choosing specific actions for any given input and rule set, the *reinforcement component* alters the parameters of the current rule set based on the environmental feedback, and the *rule discovery component* generates new, potentially superior, rules from old ones using a genetic algorithm. The authors show that the agents' behavior is broadly consistent with real-world strategies. This conclusion is grounded on the observation that nuclear units generally bid at low prices, regardless of the demand level, whereas oil/gas turbine units bid high in order to capture peak generation; coal and gas units bid close to the level required for profitability and increase their bids in times of high demand. An additional series of experiments compares uniform to pay-as-bid pricing. Results show that agents tend to bid higher under pay-as-bid, but the overall cost of meeting demand is still higher under uniform pricing. Another question or interest was the evolution of cooperation, defined as situations in which two or more players make the same high bid. Although the agents are able to produce high market prices in some environments, the authors do not find many rounds in which the cooperation criterion is met. They conclude that the number of available actions, the exploitation/experimentation policy of the LCS and the potential incorrect generalization over environments makes it difficult for agents to maintain cooperative strategies.

3.3.3 Other Agent-Based Electricity Market Simulations

Some of the first AB simulation models of electricity systems define their own representation of how agents adapt to the system they are placed in. These learning representations are usually tailored for the specific design of the simulated market(s). They do not explicitly rely on findings from psychological research about learning or on developments from the DAI or MAS fields of agent learning. These – usually naïve or intuitive formulations – are termed model-based adaptation algorithms here. The most prominent work in this field has been conducted at the London Business School; other approaches, such as those by Visudhiphan and Ilić, have also attracted interest by researchers.

Analyzing Trading Arrangements in England and Wales (Bower, Bunn et al.)

Bower and Bunn (2000) present an AB simulation model of the England and Wales electricity market. The simulation is designed to compare different market mechanisms, i.e. daily versus hourly biding and uniform versus discriminatory pricing. Generator agents apply a simple reinforcement learning algorithm which is driven by the goal to simultaneously maximize profits and reach a target utilization rate of their own power plant portfolio. The agents adjust their bidding strategies according

3.3 Related Work: ACE Electricity Market Models

to their last round's trading success; they either lower, raise, or repeat their last bid price, depending on whether their utilization and profit targets have been met in the last round, or not. The demand side of the market is modeled as a static aggregate load curve with limited price sensitivity. Transmission constraints or costs are neglected. The results from different scenarios show that simulated market clearing prices are lowest in the case of daily bidding with uniform pricing, and highest in the case of hourly bidding with discriminatory pay-as-bid settlement. The authors explain this result by two phenomena: (1) the risk of overbidding by base load generators is higher in the pay-as-bid case; this reduces competitive pressure on the mid-merit plants; (2) hourly bidding allows generators to effectively segment demand into peak load and base load hours and, thus, to extract a greater portion of the consumer surplus than under daily bidding. Another finding that the authors report from the simulation results is that for all simulated scenarios bid prices fall as the target rate of utilization rises. At a target utilization rate of 100%, prices fall to marginal cost – an observation that is consistent with theoretical considerations, i.e. that the optimal bidding strategy for a hedged power plant is to bid at marginal cost. As generator agents in the model also learn across their portfolios and can transfer successful bidding strategies from one plant to all others, small agents with few power plants have an informational disadvantage over large firms who can submit more bids and, consequently, gather more market price information. In their simulation results, the authors find that small agents perform better in relation to large agents in a uniform price setting than they do in the pay-as-bid case. This can be explained by the fact that in the uniform price case, all agents receive the information of the system marginal price, i.e. the result of the industry's collective learning; thus the informational advantage that large generators have in the case of discriminatory pricing is mitigated.

In Bower and Bunn (2001) the above described analysis is complemented by a validation of the simulation model against classical models of monopoly, duopoly, and perfect competition. The mean simulated market clearing prices for pay-as-bid and uniform pricing are very close to the corresponding theoretical results in the monopoly and perfect competition models. In the case of duopoly, however, the difference between pay-as-bid average prices and uniform system marginal prices is much smaller in the simulation model than in the theoretical benchmark. It is not clear which conclusions the authors draw from the comparison with the theoretical duopoly case and what their results imply for an oligopoly model (are prices in an oligopoly closer to the competitive equilibrium or to the duopoly case?).

Bower, Bunn, and Wattendrup (2001) apply the same basic model to the case of the German electricity sector. It is simulated as a day-ahead market in which plants are dispatched centrally and remunerated on a pay-as-bid basis. The authors analyze the impact of four mergers of large German utilities that were probable at the time of the study (and have actually taken place shortly after). They find that electricity prices rise considerably as an effect of the mergers, especially in on-peak times. When the merged firms also seek to rationalize their portfolios and shut down about part of the total system capacity, the effect on prices is even more severe, and prices raise considerably, especially during winter peaks.

Bunn and Oliveira (2001) present a more detailed model of the *New Electricity Trading Arrangements* of England and Wales. In contrast to the approach described above, the authors explicitly model an active demand side and the interactions between two different markets, i.e. the bilateral market and the balancing mechanism. Trading in both markets is modeled as a call market with pay-as-bid settlement. Both generators and suppliers seek to maximize individual daily profits and simultaneously minimize the difference between their exposure to the balancing mechanism and their (fixed) objective for balancing mechanism exposure. They learn to set mark-ups on their bid prices in both markets through reinforcement learning. All generator (supplier) agents learn three different policies for setting three different prices: the offer (bid) price for the bilateral market, and the bid prices of, respectively, *increments* and *decrements* for the balancing mechanism. In order to avoid inconsistent behavior during the learning process, the authors impose some "lower bounds of rationality" on the agents' bidding strategies through the introduction of *operational rules*. Suppliers, for example, make sure that a more flexible power plant never undercuts the offer of a less flexible plant of their same portfolio. The range of possible actions (relative mark-ups) of the reinforcement learning algorithm differ for suppliers and generators (e.g. between 0.95 and 1.2 for suppliers, and between -0.15 and 1.15 for generators in the bilateral market). At each trading day and for each possible mark-up, agents calculate the *expected daily profit*, the *expected acceptance rate*, and – as a product of these – the *expected reward*, using exponential smoothing of the previous days' trading results. Expected rewards are ranked in descending order, and *perceived utilities* are calculated on the basis of ranks of mark-ups in the order of expected rewards. The probability of choosing a mark-up is proportional to this perceived utility. The simulation model is run with power plant data that represents the UK wholesale electricity market. The authors observe very high prices in those two hours of the day in which demand is highest. Another observation is that a wide spread between the *System Buy Price* and the *System Sell Price* emerges in the balancing mechanism and that the average bilateral price is centrally located between them. This corresponds to the intuitive system behavior.

An extension of the model and further analysis is presented by Bunn and Oliveira (2003). Here, the research question to be analyzed is whether two specific generation companies in the England and Wales electricity market are capable of manipulating market prices in order to increase their profits. They compare six different withholding strategies, including the case of no withholding (benchmark), cases in which only one of the two generators withholds capacity, and one where both simultaneously withhold parts of their capacity. The reported results indicate that only one of the two generators, whose ability to influence market prices was studied, is capable of increasing electricity prices unilaterally. If both companies act together, they can also significantly increase power exchange prices. The second generator had no ability to manipulate prices alone. In order to manipulate prices profitably, both generation companies have to act together. Moreover, prices in the balancing mechanism are found to be robust against manipulation from the two players. One interesting remark by the authors is that in these types of AB models, i.e. when agents learn

3.3 Related Work: ACE Electricity Market Models

to adapt their behavior to a stable environment, the potential for agents to collude on higher than marginal costs can be overestimated. In real markets where varying demand, fuel costs, and transmission constraints change the state of the world continuously, this coordination behavior might be harder to achieve.

An agent-based analysis of technological diversification and specialization is presented in Bunn and Oliveira (2006). The question that the authors want to answer in this paper is whether strategic generator agents in the electricity markets rather evolve into diversified players with a mix of baseload, shoulder and peak load plants, or into specialized players that seek to dominate the market in their segment. They develop a model in which generators trade generating capacity among each other and then, in a second stage, trade electricity from their plants, applying a Cournot strategy. Two mechanisms are compared: single-clearing and multi-clearing. The former mechanism corresponds to a uniform price power pool; the latter is supposed to replicate trading in bilateral markets. In the multi-clearing setting, base load, shoulder, and peak load are traded separately in three different markets. Generators aim at maximizing the value of their whole portfolios. They estimate which plants are most likely to be traded and, on that basis, decide which plants they will actually attempt to buy or sell. The adaptation procedure is based on stochastic search and learning. An *inertia* element defines whether the agent searches for new strategies or stays with the old ones; it decreases over the course of a simulation. New strategies are defined as the *best response* to the given situation; this action maximizes the sum of the utility (profit/reward) of an action and the discounted portfolio value. The simulation model is run with plant data from the England and Wales electricity market. In the first set of simulations, base load, shoulder and peak plants were separated among three different players. The single-clearing mechanism leads to high concentrations in this setting: a single player (the base load player) becomes a monopolist, while the others sell all their capacity and are extinct after less than 2,000 iterations. In the multi-clearing case, several players coexist; the base-load and shoulder players own most of the capacity at equal share, whereas the peak player retains some capacity at the end of 2,000 iterations. Prices are also higher in the single-clearing than in the multi-clearing case. In a second scenario, all three players have similar initial portfolios. In this case, the difference between the two clearing mechanisms is rather small. The authors conclude from this observation that if the industry is at a state of great diversification, it will tend to remain so, independently of the market-clearing mechanism.

Comparing Different Adaptation Algorithms (Visudhiphan and Ilić)

Visudhiphan and Ilić (1999) describe a model consisting of three strategically interacting generator agents who apply some form of *Derivative Follower* strategy (Greenwald et al., 1999) for learning to set profit-maximizing bid prices. In their extremely simplified model, the authors can show that in a market with price-inelastic load, generators extensively exercise market power, while a price-responsive demand side leads generators to bid more competitively, leading to lower market prices.

In a later paper, Visudhiphan and Ilić (2001) report on simulation results from a model in which agents can strategically withhold capacity when their expected profit is higher than without withholding. Each agent records data about the market outcome in previous market rounds. The outcomes are each mapped to predefined discrete load ranges, so that each agent's memory can be represented as a matrix with rows corresponding to the different load ranges and columns corresponding to the market rounds. Bid quantities and prices are defined separately in a two-step decision process. Each agent is assigned one out of six proposed strategies for setting the bid prices of anticipated marginal units, e.g. set bid price equal the maximum or mean of historic prices, or to the sum of weighted historic prices. Simulation results are presented for two scenarios of available capacity and for strategic and competitive (i.e. marginal-cost) bidding. The authors come to the conclusion that generators are able to raise market prices if they bid strategically. This is observed not only for hours of high electricity demand, but also for low-demand hours. However, a distinction between the success of different price or quantity bidding strategies is not provided.

A more in-depth discussion of agent bidding is provided by Visudhiphan (2003). The thesis explores three different learning algorithms or bid selection strategies: (1) a modification of an algorithm formulated by Auer, Cesa-Bianchi, Freund, and Schapire (2002), (2) a simple reinforcement learning algorithm using a Boltzmann distribution for defining the probabilities of choosing each action, and (3) a model-based algorithm similar to the one presented in Visudhiphan and Ilić (2001). Agents learn bid prices and bid quantities separately through the applied learning algorithm. Various simulation runs with differing parameter combinations have been tested with these three algorithms. What is striking about the simulation results is that the daily load cycle seems to have a much stronger influence on resulting market prices than the learning representation. In all simulations, daily price cycles can clearly be distinguished, while for most of the learning algorithms prices do not exhibit any longer-term trends. The author does not provide any discussion about which learning algorithm is most appropriate for realistically modeling real-world behavior. She also concludes that her results cannot be validated against market results observed in any real-world market, because information on marginal-cost functions, bilateral contract obligations, operating constraints or other power system characteristics is not sufficiently available. Moreover, the thesis reports on comparative results from uniform versus pay-as-bid pricing; the conclusion of the result evaluation is that outcomes significantly depend on the learning algorithms that the agents employ.

Two further publications by one coauthor describe results from another simulation model for analyzing market dynamics that arise from individual agent decision making. The strategic agents are two/three generators in a transmission system with and without congestion, respectively (Ernst et al., 2004b); other runs also include a profit-maximizing transmission line owner (Ernst et al., 2004a). Here, agents strategically set the bid price that will maximize their payoff under the assumption that the other agents repeat the same actions as in the precedent trading round. This approach seems to be a renunciation from the earlier approaches presented.

Simulations with Supply Function Optimizing Agents (Day and Bunn)

While an analytical evaluation of supply function equilibria in power markets either assumes continuous supply functions or restricts the analysis of various industry ownership structures to symmetric equally sized firms (or both), Day and Bunn (2001) present a method for determining imperfectly competitive outcomes in electricity markets based on computational modeling. Their proposed model contains generation companies who bid individual piece-wise linear supply functions into a market with uniform price clearing. The agents seek to optimize the value of an objective function, i.e. their daily profits from both spot sales and long-term financial contracts. The optimization routine that the agents apply works under the conjecture that the other agents submit the same bid as in the previous trading round. Agents in this model have a limited optimizing behavior in that they only change the bid price of one power plant per iteration (they choose the plant which increases the value of the objective function most when bid at another price). Electricity demand is represented by an aggregate demand function with a defined demand elasticity; simulations are run with different demand elasticities in order to determine the influence of demand response on the generators' ability to exercise market power. The authors evaluate their computational model by comparing it with the equilibrium in continuous supply functions that can be obtained through the approach formulated by Klemperer and Meyer (1989). They find that results from the two approaches are reassuringly close for a simplified market scenario which models competition between three symmetric generating companies with linear marginal costs. Based on this finding, the authors are confident that the computational approach can also deliver realistic results for more complex scenarios that cannot be represented by an analytical supply function equilibrium model. They use the computational model for analyzing different options for the second round of plant divestiture in the England and Wales electricity market. Results from various runs with different demand elasticities and different volumes of financial forward contracts show that the analyzed divestiture options result in lower average percentage bids above marginal cost. One result that the authors point out is that the main reduction in mark-up is caused by the creation of five generators (from initially three); the difference in divestiture percentage has only a small effect on observed mark-ups. However, the authors also find that the proposed divestiture still leaves considerable market power with generators in the short term, and could result in prices above short-run marginal costs.

In a later paper Bunn and Day (2002) present this model as a competitive benchmark against which to assess generator conduct and to diagnose the separate causes of market structure and market conduct in situations where prices appear to be above marginal costs. They argue that the result from their computational model can serve as a realistic baseline for imperfect (oligopoly) competition, where agents learn to compete, but not to collude. For the tested scenarios, the simulated system supply functions are shown to lie above the marginal cost function and significantly below the system supply curve observed in the England and Wales pool on an exemplary day, except at low demand levels. This leads the authors to the conclusion that the extent to which the simulated supply functions are above the marginal cost function is caused by the market structure. The extent to which observed system supply functions in the real-world market are still above the simulated system supply functions is then interpreted as the degree of collusion within the market, and identifies a problem of market conduct.

Large-Scale National Agent-Based Electricity Simulations

To the author's knowledge, three national U.S. and one Australian laboratories are currently developing large-scale AB electricity system models which are intended to serve as tools for reliability and market design analysis for power markets:

- *EMCAS*, the *Electricity Market Complex Adaptive System* developed by Argonne National Laboratory (Conzelmann et al., 2005)
- *Marketecture* from Los Alamos National Laboratory (Atkins et al., 2004)
- *N-ABLE^{TM}*, the *Agent-Based Laboratory for Economics* developed at Sandia National Laboratory (Ehlen & Scholand, 2005)
- *NEMSIM*, the Australian *National Electricity Market Simulator* which is under development at CSIRO (Batten & Grozev, 2006)

These models rely on very detailed databases of the regions under study, which includes the topology of the transmission grid and other physical constraints, differentiated load data and detailed cost data of the power plants. The issues that most of the models address are questions of best market designs to prevent the exercise of market power, to ensure transmission system reliability, or also for environmental regulation measures. The models seem to be designed as a decision support for concrete policy making and not primarily for academic research. Consequently, the model descriptions are rather vague and much of the exact implementation of agent behavior or simulated scenarios remains unclear.

Some of the cited papers contain pointers to the literature about learning algorithms though no actual model implementation using any kind of agent learning is presented. The agents' behavior has been described in more detail for two models: In one scenario simulated with the *N-ABLE^{TM}* model, agents apply a heuristic planning process in the form of a *greedy* scheduling algorithm. The research question of this scenario was the influence of real-time pricing contracts on consumption and profitability in the retail electricity market (Ehlen et al., 2007). In the *Marketecture* model agents follow one out of three possible fixed strategies: they set bid prices and quantities according to the *competitor*, *oligopolist*, or the *competitive-oligopolist* strategy. The *competitor* strategy refers to bidding at marginal cost, whereas the *oligopolist* bids at the point where the marginal revenue and marginal cost functions intersect; the *competitive-oligopolist* strategy lies at a random point in the range between the two other strategies. Buyers' and sellers' surplus, efficiency and market clearing prices/quantities are compared for three different market clearing algorithms (Atkins et al., 2004). However, as agents do not adapt to the different market clearing rules, no well-grounded conclusions can be drawn on the efficiency of these rules. In summary, the ability of the cited large-scale AB models cannot be judged here, as too few simulation results have been published.

3.3.4 Discussion of ACE Electricity Approaches

In order to allow a better comparison and a brief overview of the different modeling approaches, the presented papers are shortly summarized in Table 3.1. If the same authors have described similar or enhanced models in several papers, only the most relevant reference is listed. As for their rather practical application and not primarily academic focus, the large-scale simulation models are not included in this overview. The comparison of the different model shows the similarities and differences between current AB electricity models:

- The large majority of models neglect transmission grid constraints.
- Most of the models represent the demand side as a fixed, price-insensitive load.
- In most models, the agents' learning task is to set profit-maximizing bid prices or mark-ups. Capacity withholding strategies are mostly not modeled explicitly; however, setting a high bid price can also be interpreted as (economic) withholding.
- No preferred learning representation or trend towards specific models of behavior can be observed. A number of models rely on the reinforcement learning algorithm formulated by Erev and Roth (1998), and on its modifications; however, they do not form a considerable majority. Genetic algorithms seem to be left apart, though not completely abandoned.
- Most research questions of AB modelers center around market power and market mechanisms. The comparison between pay-as-bid and uniform pricing is a very popular question. Here, results from different models seem to be consistent; most authors find that agents bid higher under pay-as-bid pricing, but overall prices are higher under uniform pricing. Another important research issue for AB electricity modelers is the assessment of the market power potential under different market structures or market mechanisms.

The amount of papers reviewed in this survey shows that electricity market research applying AB simulation is a very active field of research. One might describe it as adolescent – it has departed from its infancy which began in the last years of the past century. This is documented by the appearance of first notable papers that have successfully been published in energy-related or other journals. However, we still observe a large heterogeneity in representing boundedly rational actors in electricity markets, and also in validation techniques, result evaluation and quality assessment, or simply in labeling. Agent-based modeling allows for great flexibility in specifying how agents behave; the reverse of this medal is that models are rarely comparable, and can sometimes not be described in all necessary detail. On its way to adulthood, hence, several methodological questions will have to be discussed by researchers who are active in this field, in order to ameliorate the comparability of different models. Some of these issues are enumerated in the following.

Table 3.1 Summarized overview of agent-based electricity market modeling approaches

References	Agents' actions	Market	Demand side	Transmission system	Research questions
Erev and Roth reinforcement learning (Sect. 3.3.1)					
Nicolaisen et al. (2001)	Set bid prices	Double-auction with discriminatory midpoint pricing	Adaptive demand side (both buyers and sellers of electricity bid in the auction)	No transmission constraints	Analysis of buyers and sellers market power under different concentration conditions; distinction between structural market power and market power due to agent learning
Sun and Tesfatsion (2007)	Set supply functions	Real-time market, day-ahead market with locational marginal pricing	Fixed inelastic demand	Bid-based DC optimal power flow problem	Market design reliability
Rupérez Micola et al. (2006)	Set bid prices on three subsequent markets	Wholesale and retail electricity market, natural gas market	Fixed inelastic demand	No transmission constraints	Can agents benefit from vertical integration? How can agents exert vertical market power?
Bin et al. (2004)	Set bid prices	Call market with uniform, pay-as-bid and *electricity value equivalent* pricing	Fixed inelastic demand	No transmission constraints	Comparison of resulting prices for the three pricing mechanisms
Cincotti et al. (2005); Cincotti, Guerci, Ivaldi, and Raberto (2006)	Set price quantity bid pairs	Call market	Fixed inelastic demand	No transmission constraints	Comparison of bidding strategies and resulting prices in pay-as-bid and uniform price auctions
Weidlich and Veit (2006)	Set price quantity bid pairs	Call market, procurement auction for balancing power	Fixed inelastic demand	No transmission constraints	Dynamics between two interrelated markets, comparison of prices in pay-as-bid and uniform price settlement cases
Q-learning (Sect. 3.3.1)					
Krause et al. (2005); Krause and Andersson (2006)	Set bid prices	Generation cost minimizing ISO (considers network constraints)	Fixed demand; linear function of demand side marginal benefit	DC network representation with transmission capacity constraints	Comparison between Q-learning and Nash equilibrium strategies; evaluation of different congestion management mechanisms

3.3 Related Work: ACE Electricity Market Models

Reference	Agent action	Market	Demand	Network	Research question
Naghibi-Sistani et al. (2006)	Set slope of linear bid function	Uniform price call market	Fixed price-elastic demand	No transmission constraints	Comparison between Nash equilibria and the proposed Q-learning algorithm with temperature variation
Xiong et al. (2004)	Set bid prices	Uniform price and pay-as-bid call market	Both fixed inelastic demand and demand response	No transmission constraints	Comparison of market prices in uniform price and pay-as-bid auctions
Bakirtzis and Tellidou (2006)	Set bid prices	Uniform price and pay-as-bid call market	Fixed inelastic demand	No transmission constraints	Comparison of market prices in uniform price and pay-as-bid auctions
Richter and Sheblé (1998); Petrov and Sheblé (2000)	Set bid prices	Double-auction with uniform pricing	Genetic algorithms (Sect. 3.3.2) Fixed inelastic demand	No transmission constraints	Examination of bidding strategies
Nicolaisen et al. (2000); Lane et al. (2000)	Set bid prices	Double-auction with uniform pricing	Active demand side bidding	No transmission constraints	Measure market power exerted in a double-auction
Cau and Anderson (2002); Cau (2003)	Assign bid quantities to price segments	Cost minimizing ISO	Inelastic, uncertain demand (*high/low* with equal probability)	No transmission constraints	Analysis of collusive strategies by the agents
			Learning classifier systems (Sect. 3.3.2)		
Bagnall (2000a); Bagnall and Smith (2005)	Set bid prices	ISO who integrates unit commitment constraints into the allocation calculation	Fixed half-hourly forecast demand	No transmission constraints	Can the agents evolve behaviors observable in the real world? How do market mechanisms (uniform price *vs.* pay-as-bid) effect bidding behavior?
		Supply function optimization heuristic (Sect. 3.3.3)			
Day and Bunn (2001); Bunn and Day (2002)	Assign bid quantities to price segments	ISO calculating the system marginal price	Fixed, price-elastic demand (locally linear functions)	No transmission constraints	Differentiation between market structure and market conduct as a reason for high electricity prices; impact of divestiture proposals on the players' potential to exert market power

(continued)

Table 3.1 (continued)

References	Agents' actions	Market	Demand side	Transmission system	Research questions
Model-based learning algorithms (Sect. 3.3.3)					
Bower and Bunn (2000, 2001)	Set bid prices for portfolio of plants	Day-ahead market with uniform or pay-as-bid clearing	Static price responsive demand	No transmission constraints	Comparison of pay-as-bid vs. uniform price clearing, and daily vs. hourly bidding (impact on prices)
Bower et al. (2001)	Set bid prices for portfolio of plants	Day-ahead market with uniform or pay-as-bid clearing	Static price responsive demand	No transmission constraints	Impact of merger options in the German electricity market on wholesale prices
Bunn and Oliveira (2001, 2003)	Set mark-ups on bid prices in both markets separately	Forward market with pay-as-bid clearing, balancing mechanism	Active demand side bidding	No transmission constraints	Evaluation of generators' conduct in the England and Wales market: can either one of the generators, or two of them together exercise market power?
Bunn and Oliveira (2006)	Trade power plants with other agents; play Cournot strategy in the electricity market	Market for power plant capacity; day-ahead electricity market	Static hourly linear demand functions	No transmission constraints	Do electricity markets tend towards technological diversification or specialization? Which influence does the market clearing have on this question?
Visudhiphan and Ilić (2002)	Set bid prices and apply capacity withholding; maintenance and investment decisions	Welfare maximizing ISO	Inelastic demand function with uncertainty	No transmission constraints	Assessing electricity market efficiency and market power exertion over short, medium and long term horizons (no simulation results, only "work in progress")
Visudhiphan and Ilić (2001); Visudhiphan (2003)	Set bid prices and quantities (step-wise bid functions)	Uniform price call market	Fixed inelastic demand	No transmission constraints	Realistic representation of market price dynamics and participants' bidding behavior in electricity markets; role of generator learning for strategic bidding

3.3 Related Work: ACE Electricity Market Models

Agent Learning Behavior

The common element of all models presented here is adaptivity. Agents are able to learn to achieve their goals (high profits, high plant utilization, etc.) given the environment they are placed in. The way in which learning takes place is implemented differently in almost all models. Even if two researchers use the same basic learning algorithm, they may define and describe it in a different manner. Parameter combinations are often not justified properly and not fully revealed. A precise description of an agent's action domain or the space of possible environmental states is not defined clearly in every paper. Especially in the models applying Q-learning, the definition of environmental states has implications on the model results; however, none of the cited papers using Q-learning argues why the states have been set as they are, and whether they have the Markov property. Moreover, the choice of the learning algorithm itself is hardly argued and justified in any paper. Some algorithms like, e.g. the Erev and Roth reinforcement learning formulation, are popular, others are hardly used at all, without apparent reason.[21] Most papers do not answer the question why they chose a specific learning model and how good it performs in comparison to others. Also, it might be interesting to discuss if there is any meaningful minimum level of rationality that agents participating in electricity markets should be endowed with.

Market Dynamics and Complexity

As stated in the introduction, the electricity sector is characterized by the interlinking of multiple markets, and by additional complexities through limited transmission capacities and unit commitment constraints. Most papers considered here simplify real-world markets significantly; they consider only one market or neglect technical constraints. Some AB models are so highly stylized that they cannot claim to be more realistic than traditional equilibrium models. Also, the focus of most researchers is placed on convergence towards stable market outcomes. The out-of-equilibrium dynamics or the way towards an equilibrium are not considered. It might also be interesting to examine under which circumstances agents reach an equilibrium outcome and when they fail to do so; or, in case of multiple equilibria, which outcome occurs more frequently and how robust these outcomes are against some changes in parameter values. Usually the results of an agent-based simulation run are boiled down to one convergence price per simulation. The characteristics of the time series of prices are usually not examined in any detail (in order to answer questions like "How volatile are prices?", "Can price spikes be observed?", "Are prices mean reverting?", etc.). Also, the agents' profits and success of different trading strategies are rarely discussed. One important aspect of the electricity sector that can

---

[21] To give an example for a learning representation that has hardly gained any attention by AB electricity researchers, one might mention Experience-Weighted Attraction; see Sect. 4.1.3 for a discussion of this learning model.

perhaps best be represented in AB models is not considered in any of the presented papers: bilateral trading. It might be interesting to compare the efficiency of market outcomes in a bilateral setting with that of a centralized auction. Moreover, vertical integration can realistically be modeled in agent-based simulations. With few exceptions, however, this aspect is neglected in current agent-based electricity sector models. We would suggest that these still neglected factors should be stressed more in future modeling approaches.

Calibration and Validation

Calibrating and validating agent-based electricity models is a challenging task, and only few guidelines for this process have yet been defined. Many of the reviewed papers in this survey lack information about empirical model validation. Those few researchers who report about the (empirical) validation of their model (Macal & North, 2005 is one example) proceeded in heterogeneous ways. Very recently, the need for reliable validation techniques has obviously been recognized. AB researchers have analyzed and suggested procedures for calibrating and validating agent-based simulation models. Their general suggestions should now be assessed from the perspective of their usefulness for electricity modeling purposes. The development of guidelines for assuring the validity of AB electricity models would greatly benefit the research quality and diminish the heterogeneity of approaches in this field. It should thus be one of the main tasks for future work.

Model Description and Publication

Just as model verification and validation is done very differently in the cited papers, so is the description of the model, its parameters, and the results. Some papers do not deliver information about the number of runs they have conducted, some do not even publish all model parameters so that it is not possible to replicate the reported results. It would be helpful if some standard way of model description, as it is conventional for other economic methodologies, became accepted in the medium-term. Presumably, many models are still used by their developers, so that these are reluctant to make their source code available. However, this would greatly benefit the research field, because researchers could revise and check the implementations of others and could also reuse parts of them. Prof. Tesfatsion has taken the initiative into this direction by setting up a websites with links to published sets of AB electricity model source code[22] (and by publishing her source code as well).

---

[22] http://www.econ.iastate.edu/tesfatsi/ElectricOSS.htm

3.4 Summary

In this chapter, the methodology of Agent-Based Computational Economics has been described. A general dissatisfaction with highly stylized analytical economic models that often have to rely on unrealistic assumptions in order to be tractable has led to the increased use of computational simulations that allow more flexibility in model specification. The process of building ACE models is still heterogeneous; however, basic concepts, some validation techniques and helpful software tools have been presented in this chapter. Furthermore, this chapter has critically reviewed a considerable amount of relevant papers in agent-based electricity market research. Table 3.1 summarizes the core characteristics of the cited work and displays the similarities and differences between the approaches. It has identified some of the current problems facing this research methodology that require further effort and consolidation. Especially, sound argumentations for the choice of specific learning algorithms, more careful and well documented validation and verification procedures as well as the appropriate publication of details of concrete simulation models are crucial for the further development of AB electricity market modeling.

Despite the open issues and problems, AB electricity research has been successful in recent time. Many researchers have managed to replicate core characteristics of today's electricity markets using models with adaptive, self-seeking agents. With a decrease of heterogeneity between competing models, and with increasing consensus on important methodological questions, the field of AB modeling can soon become one major strand of research for the analysis of complex electricity systems.

Part II
An Agent-Based Simulation Model for Interrelated Electricity Markets

Chapter 4
Representation of Learning and Adaptation

The agent-based electricity sector simulation model that will be fully described in Chap. 5 comprises adaptive agents that bid in interrelated electricity and CO_2 allowance markets. In this model, the representation of strategic bidding behavior is a central element, and is realized through *learning algorithms*.

The review of existing agent-based approaches to simulating electricity markets (see Sect. 3.3) revealed that numerous different models of adaptive and learning behavior have been applied by ACE researchers, most of them without clear justification, let alone comparison of several alternatives that determine the most appropriate model or the best parameter combination. This observation can be made for models involving learning in other economic contexts, too (Brenner, 2006). Meanwhile, the representation of learning constitutes the core element of agent-based simulations and much effort should be spent for justifying the chosen approach. If several models of learning or adaptation are suitable, they may all be used, thereby possibly rejecting those conclusions that are not confirmed by all applied learning models, thus making confirmed simulation results better grounded.

Following the aforementioned argument, this chapter reviews the candidates for behavioral representations in agent-based electricity market models. Three basic concepts of behavioral models have attracted attention by economic researchers (Duffy, 2006): *zero-intelligence* approaches, *reinforcement* and *belief-based* models, and *evolutionary* concepts. The last two learning models will be presented in Sects. 4.1 and 4.2. This presentation is followed by the description of a set of simulation runs that have been carried out in order to decide which learning algorithm best represents agents' bidding behavior in a simplified electricity market scenario (Sect. 4.3). A number of different variants and parameter combinations of three learning models have been tested and compared extensively. They can be categorized as reinforcement learning (RL) and hybrid reinforcement and belief-based learning algorithms.

4.1 Reinforcement and Belief-Based Learning

Selten (1991) mentions three kinds of learning: *rote learning*,[1] *imitation*, and *belief learning*. While imitation plays a more important role in evolutionary learning models (see Sect. 4.2), the other two learning classes have been expressed in different learning algorithms that will be discussed in this section.

In many studies analyzing the predictive accuracy of reinforcement or belief-based learning algorithms, it has been shown that these are better predictors of human subject behavior than static Nash equilibrium point predictions. This observation constitutes one central motivation of agent-based approaches in economics. If reinforcement or belief-based models of learning represent human behavior in many strategic situations more realistically than static equilibrium calculations, these models are well suited for the analysis of various economic problems. Besides, computational learning algorithms are more generic than equilibrium calculations. Thus, they can also be applied to more complex scenarios, because they place less restrictions on the formulation of the model they are used for. Hence, they are appropriate for the agent-based simulation model of interrelated electricity markets, which is formulated in this work.

At least two distinct areas of research have proposed reinforcement learning algorithms. The first one is the *psychological* learning research field. Here, the question of how good an algorithm imitates human behavior, that is how close learning outcomes from the algorithm match those of human subjects, is in the focus of analysis. First propositions of algorithms that formalize human learning behavior have been given by Bush and Mosteller (1955); an early application of reinforcement learning to economics has been conducted by Arthur (1991). Another example for an RL algorithm developed from the perspective of psychological findings is discussed in more detail in Sect. 4.1.1 (Erev & Roth, 1998).

The other field of research in which reinforcement learning plays a significant role is *computer science*, especially its *machine learning* and *Multi-Agent System* streams. In this area, convergence to optimality is an important aim, and realistic modeling of human behavior is not. Q-learning (see Sect. 4.1.2) is maybe the most popular example for machine learning algorithms using reinforcement learning principles. Others are surveyed by Sutton and Barto (1998) and Kaelbling, Littman, and Moore (1996).

Reinforcement learning is basically trial-and-error learning. An agent faces the same situation repeatedly and tries to discover which actions yield the best outcomes. Kaelbling et al. (1996) define RL as follows: "Reinforcement learning is the problem faced by an agent that must learn behavior through trial-and-error interactions with the dynamic environment". To this definition, it should be added that the learner, i.e. the agent, is goal-oriented (e.g. seeks to maximize profit) and continually tries to find the best actions for achieving his goal. In doing so, he interacts with his environment and receives feedback from it.

---

[1] Rote learning = synonym for reinforcement learning.

4.1 Reinforcement and Belief-Based Learning

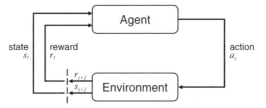

Fig. 4.1 Agent–environment interaction in reinforcement learning (Sutton & Barto, 1998)

A simple scheme of the iterative agent–environment interaction is given in Fig. 4.1: departing from state s_t and the experience from the last chosen action, the agent chooses an action a_t at time step t. This action yields a reward r_{t+1}, e.g. profit, or any other numerical value that the agent seeks to maximize over time. Simultaneously, the agent's state has changed to s_{t+1} as a consequence of the chosen action.[2] The agent then evaluates the result achieved and changes the probabilities of choosing each possible action (differentiated by states) based on his experience. The procedure how probabilities are updated differ across RL algorithms.

In reinforcement learning, agents do not form beliefs about the characteristics of the game they are playing or the market they participate in. They need not even know that they are playing a game or bidding in a market. In contrast, with belief-based learning models, agents recognize that they are playing a game against other opponents or that they participate in a market with competitors. They form beliefs about the likely play of these opponents or competitors, and adapt their own strategies according to these beliefs. Beliefs are formed on the basis of the history of the other agents' chosen actions. It is assumed that the history of the game provides information about the future. In the simplest belief-based learning model, the *best-response rule*, also referred to as *Cournot rule*, simply assumes that the other players will choose the same action as in the previous round. An agent responds in the best way if he myopically maximizes his expected payoff. The *fictitious play rule* looks further back into the past. In this learning model, agents assume they are facing a stationary but unknown distribution of opponents' strategies. They iteratively learn this distribution by equating current beliefs to the average of all previously observed action distributions (Fudenberg & Levine, 1998).

In the following, three learning models will be presented in more detail: two RL algorithms – Erev and Roth reinforcement learning and Q-learning – and one that combines aspects from reinforcement learning and belief-based learning – Experience-Weighted Attraction.

---

[2] Through the *state* variable, both the agent's internal and the environmental state can be expressed. Some RL algorithms do not differentiate between different environmental states. These are called *stateless* learning algorithms.

4.1.1 Erev and Roth Reinforcement Learning

Based on psychological findings about human learning, Erev and Roth (1998) have developed a three parameter reinforcement learning algorithm. In developing their algorithm, they departed from two basic principles that are described in the psychological learning literature. These are the *Law of Effect* and the *Power Law of Practice*. The former has first been described by Thorndike (1898) and says that choice behavior is probabilistic and choices that led to good outcomes in the past will be chosen again with a higher probability than less successful choices. The latter principle hearkens back to Blackburn (1936) and describes the observation that learning is stronger at the beginning (described by steep learning curves) and becomes less important after some time of training (flat learning curves).

The learning formulation integrates the aforementioned aspects through the introduction of an *experimentation* parameter and a *forgetting* parameter. Every possible action that an agent can choose is assigned a *propensity* that determines the likelihood that this action is chosen in further rounds. The initial propensity values $q_{ij}(0)$ for each agent i and action j are assumed to be equal for all possible actions, i.e. agents have no prior experience of the game they play. Propensities are updated at every iteration in the following way:

$$q_{ij}(t+1) = \begin{cases} (1-\phi)q_{ij}(t) + R(x)(1-\varepsilon) & \text{if } j = k \\ (1-\phi)q_{ij}(t) + R(x)\frac{\varepsilon}{M-1} & \text{if } j \neq k \end{cases} \quad (4.1)$$

Here, q_{ij} corresponds to the propensity of agent i to choose action j, $R(x)$ is the reinforcement based on the payoff x achieved from the last chosen action k, M denotes the total number of possible actions, ϕ is the recency (or forgetting) parameter and ε is the experimentation parameter.

In games in which similar strategies can be linearly ordered – as it is the case for ordered bid prices or quantities – Erev and Roth propose to "spill over" some part of the reinforcement to the neighboring (similar) actions:

$$q_{ij}(t+1) = \begin{cases} (1-\phi)q_{ij}(t) + R(x)(1-\varepsilon) & \text{if } j = k \\ (1-\phi)q_{ij}(t) + R(x)\frac{\varepsilon}{2} & \text{if } j = k \pm 1 \\ (1-\phi)q_{ij}(t) & \text{otherwise} \end{cases} \quad (4.2)$$

In the simplified scenario that will be presented in this chapter, a two-dimensional action domain, consisting of a range of possible prices and quantities, is applied. The procedure of spilling over part of the reinforcement to neighboring actions in a two-dimensional action domain is presented in Appendix A.1.

Petrov and Sheblé (2001) and Nicolaisen et al. (2001) describe one problematic feature of the original algorithm formulation, i.e. that no propensity update occurs when profits are zero (or close to zero). Another flaw of the original algorithm formulation is accentuated by Koesrindartoto (2002): for some parameter combinations of ε and M, no learning occurs. These considerations lead to the formulation of the *Modified Roth and Erev* algorithm (MRE, Nicolaisen et al., 2001):

4.1 Reinforcement and Belief-Based Learning

$$q_{ij}(t+1) = \begin{cases} (1-\phi)q_{ij}(t) + R(x)(1-\varepsilon) & \text{if } j = k \\ (1-\phi)q_{ij}(t) + q_{ij}(t)\frac{\varepsilon}{M-1} & \text{if } j \neq k \end{cases} \quad (4.3)$$

The actual choice of the action to be taken in the next round is probabilistic, with choice probabilities p_{ij} derived from the actions' propensities q_{ij}. In their original paper, Erev and Roth propose a *proportional action selection rule* as described by the following formula:

$$p_{ij}(t) = \frac{q_{ij}(t)}{\sum_{k=1}^{M} q_{ik}(t)} \quad (4.4)$$

If this action selection strategy is used, it has to be assured that propensities do not take negative values, because this would result in negative probabilities, which cannot be interpreted sensibly. In order to ensure that propensities are always positive, Erev and Roth define the reinforcement as

$$R(x) = x - x_{min} \quad (4.5)$$

where x_{min} is the smallest possible payoff.

Other researchers also use action selection probabilities defined by a Gibbs–Boltzmann distribution with a positive *temperature*, or *cooling* parameter τ. In accordance with the terminology used by Sutton and Barto (1998) this rule for determining probabilities is called *Softmax action selection rule* and is defined in (4.6). With Softmax action selection, payoffs are admitted to adopt negative values.

$$p_{ij}(t) = \frac{e^{q_{ij}(t)/\tau}}{\sum_{k=1}^{M} e^{q_{ik}(t)/\tau}} \quad (4.6)$$

The temperature parameter determines the degree to which the agent focuses on actions with high propensity values. Low temperatures τ lead to a greater difference in the action selection probabilities when actions have different Q-values, whereas high temperatures cause all actions to have nearly the same probabilities. Usually, temperature is decreased over the course of a simulation, in order to allow more exploration at the beginning, while focusing on exploitation later on. In other papers describing learning models, also the inverse value of the temperature is used, which is then called the *focus* parameter (e.g. Camerer & Ho, 1999). This notation is also adopted within this work.

Erev and Roth could show that their proposed learning model predicts data from twelve different experimental games better than static equilibrium predictions, and also better than other learning models that the same authors have developed. Their algorithm has gained much attention by agent-based modelers. Also, a considerable part of AB electricity market models apply this learning algorithm, or its modification. For this reason, it is analyzed in detail in the simulation section (Sect. 4.3) of this chapter.

4.1.2 Q-Learning

Q-learning is another reinforcement learning algorithm which has been proposed by Watkins (1989). In contrast to the Erev and Roth algorithm, a Q-learning agent not only evaluates the consequences of an action in terms of immediate reward, but he also estimates the value of the new state to which the action has taken him. The algorithm works by estimating the values of state-action pairs. The value $Q(s,a)$ is defined as the expected discounted sum of future rewards obtained by taking action a from state s and following an optimal policy thereafter. A *policy* in this context is a mapping from perceived states of the environment to actions to be taken when in those states (Sutton & Barto, 1998). Q-values are iteratively updated on the basis of the experience gained, according to the Q-learning rule:

$$Q(s_t, a_t) \leftarrow Q(s_t, a_t) + \alpha \left[r_{t+1} + \gamma \max_a Q(s_{t+1}, a) - Q(s_t, a_t) \right] \quad (4.7)$$

where $0 < \gamma < 1$ is the discount factor of future rewards and $0 \leq \alpha < 1$ is the *learning rate*. It can be shown that $Q(s,a)$ converges to the real value of an optimal policy rule $Q^*(s,a)$ with probability one in a stationary environment, if the state-action-pairs are updated infinitely and if the probabilities of the transition from state s_t to state s_{t+1} are *Markovian*, i.e. they depend only on the current state and the action taken, and not on previous states.

Several possible rules for action selection are used in practice. As for all reinforcement learning algorithms, action selection rules must ensure both *exploration* in order to let the agent discover the value of seldom chosen actions, and *exploitation* of good strategies. The ε-greedy action selection rule is one frequently used compromise between exploitation and exploration. It says that the action with the highest Q value is chosen with probability $(1-\varepsilon)$, and a random action with probability ε. This rule guarantees that all actions are continued to be played, and so the estimation of their Q-value continues to be updated. The disadvantages of this rule, however, is that when exploring, very bad actions have the same probability of being chosen as those actions that are close to the best one. Here, the Softmax action selection rule (4.6) offers a way to make the probability of an action to be chosen a function of its Q-value. Hence, Softmax action selection is also often applied with Q-learning.

4.1.3 Experience-Weighted Attraction

A learning model that integrates the agents' belief about the reward they could have achieved if they had chosen a different action has been formulated by Camerer and Ho (1999). Their Experience-Weighted Attraction (EWA) model is a formulation that includes both reinforcement learning and belief-based *weighted fictious play* as special cases.

4.2 Evolutionary Learning Models

Two variables are updated after each round in the EWA model. These are the *attractions* A_{ij} of action j for agent i and the *weight* or *number of observation equivalents* $N(t)$. The latter variable can be interpreted as the number of periods of actual experience, which is equivalent in attraction impact to the pre-game thinking; it is updated after every round as follows:

$$N(t) = \rho \cdot N(t-1) + 1, \ t \geq 1 \tag{4.8}$$

Here, ρ denotes the *retrospective discount factor* that depreciates previous experience. The update rule for the attractions considers the payoff achieved in the previous round, but also payoffs that unchosen actions would have yielded. *Hypothetical payoffs from unchosen actions are weighted by a parameter δ, and actual payoffs from action k by an additional $(1-\delta)$*:

$$A_{ij}(t+1) = \begin{cases} \frac{\phi \cdot N(t) \cdot A_{ij}(t) + \pi_i(a_{ij}, a_{-i}(t+1))}{N(t+1)} & \text{if } j = k \\ \\ \frac{\phi \cdot N(t) \cdot A_{ij}(t) + \delta \cdot \pi_i(a_{ij}, a_{-i}(t+1))}{N(t+1)} & \text{if } j \neq k \end{cases} \tag{4.9}$$

The attraction $A_{ij}(t)$ of action j perceived by agent i constitutes the sum of a depreciated, experience-weighted previous attraction plus the (weighted) payoff π from period t, normalized by the updated experience weight. ϕ is a decay rate which discounts previous attractions; a_{ij} are the possible actions of agent i and a_{-i} is the action combination of all agents except i. The probability of choosing an action j in the EWA model is also determined by a Gibbs–Boltzmann distribution (i.e. Softmax action selection, see (4.6)).

EWA has been tested against empirical outcomes of different games in several experiments (e.g. Arifovic & Ledyard, 2004; Camerer & Ho, 1999). It has been shown that the learning model performs well in simulating the empirically observed learning outcomes.

4.2 Evolutionary Learning Models

Another popular learning paradigm which can be used to model both individual learning (learning from individual experience or beliefs) and also social learning (learning by observing others agents) are evolutionary algorithms. Evolutionary learning models are especially suitable when the domain of possible actions or strategies is very large, because new strategies are continued to be generated through the process of evolution. The idea of applying processes known from biological evolution to simulations involving adaptive agents can be traced back to Holland (1975). Examples for early applications of evolutionary algorithms to economic models have been demonstrated by Axelrod (1987) and Miller (1986); they simulated the Iterated Prisoners' Dilemma with the help of genetic algorithms.

Evolutionary algorithms apply the idea of natural selection or *survival of the fittest* to computational learning or adaptation models. The fitness of a strategy (action decision) is evaluated on population-based measures. Some evolutionary concepts apply further biological principles like crossover and mutation for altering the population of strategies. Among the various evolutionary learning models that have been developed, three are frequently applied to various economic problems:

- *Replicator Dynamics* is a very simple class of evolutionary algorithms and can be applied for games with small strategy spaces. The population of strategies does not evolve, but the proportion of strategies with higher than average fitness increases over time, while worse than average strategies die out.
- *Genetic Algorithms* (GA) is a class of heuristic search methods which imitates the biological process of evolution. Strategies are encoded into bitstrings which can be thought of as *chromosomes*. Early GAs used the binary $\{0,1\}$ alphabet for coding strategies; more recent implementations also use other encodings for strategy description. The most successful (or "fittest") strategies are passed from one generation to the next by a selection and mating process in which *parent* chromosomes produce *offsprings*. By mimicking *crossover* and *mutation*, GAs exploit the genetic dynamics underlying natural evolution to succeed in their environment. Crossover is the process of paring two strings to one new string by cutting both parent strings at one or more points and swapping all data beyond one point or between two points between the two parent strings. Mutation means that single elements of the bitstring are changed randomly with a small probability; for example in a binary string, a 0 is flipped to 1 or vice versa. When applied to economic systems, the individual chromosomes of a GA population can either be interpreted as individual agents, or as one strategy that a single agent might adopt. In both cases, the interpretation of what selection, crossover and mutation corresponds to in respect to the agents' learning process differs. If a population of heterogeneous agents is to be modeled (e.g. as in electricity market simulations, agents with different generating costs), individual strategy populations for every agent are necessary, because preferable and successful actions differ between agents (Marks, 2006). Examples of agent-based electricity market models applying GA are given by Cau (2003), Cau and Anderson (2002), or Nicolaisen et al. (2000).
- *Learning Classifier Systems* (LCS) combine reinforcement learning (for increasing the probability of choosing successful actions) and genetic algorithms (for producing new rules from successful ones). A classifier is a rule of the form "if <condition> then <action>", where the condition describes the state of the environment. An LCS is thus suitable for agents to solve cognitive tasks. Examples of agent-based electricity market models applying LCS are given by Bagnall and Smith (2005) or Bagnall (2000a).

Evolutionary concepts are inspired by biological processes and are not designed to mimic psychological aspects of learning. This makes them difficult to interpret and also difficult to test for goodness of fit against human subject experiments.

Besides, they assume that fitness values across the whole population of strategies are always available for comparison purposes; if learning takes place on the level of the agent population, this approach can be regarded as less decentralized or less "agent-based" as reinforcement learning or belief-based learning (Duffy, 2006).

If the more complicated evolutionary algorithms are to be employed in agent-based economic simulations, than because they perform better than simpler machine learning alternatives. Q-learning, as one representative of machine learning algorithms, is a rather simple but effective approach to representing human learning behavior. More complicated approaches should only be applied if these simpler concepts are not sufficiently flexible or powerful to model agent learning processes. Rapoport, Seale, and Winter (2000) formulate this postulate felicitously: "[...] the simplest model should be tried first, and [..] models postulating higher levels of cognitive sophistication should only be employed as the first ones fail". Consequently, evolutionary algorithms will only be considered as a method for representing agents' behavior in this work if reinforcement or belief-based learning do not deliver satisfying simulation results.

4.3 Analysis of Learning Algorithms for Agent-Based Simulations

With the presented options of representing learning processes in agent-based simulations at hand, the question which variants of these learning models are most appropriate for simulating the agents' behavior in the electricity market model developed in this work should now be analyzed with the help of a simplified market scenario. First, the criteria for appropriate learning algorithms are formulated in Sect. 4.3.1. The simplified market scenario with which the different learning models, their variants and parameter value combinations are tested is specified in Sect. 4.3.2, and results from these simulations are presented in Sect. 4.3.3. In the conclusion drawn from this analysis (Sect. 4.3.4), the learning representation for further more detailed and more realistic simulations of the German electricity market is decided.

4.3.1 Criteria for Choosing a Learning Model

The aim of every modeling exercise is, of course, that the system under study is represented as realistically as possible, at least for those aspects that are important to the analysis that the modeler aims at. From a practical perspective, the model should also be as simple as possible, yet accounting for all necessary details. While some researchers may deem it important that the details of how agents learn or adapt are correctly represented, for most ACE modeling purposes it is more relevant that the aggregate outcomes that emerge from repeated interactions among learning agents correspond to some known stylized facts. In the analysis presented here, the minimum requirements for realistic learning representations are also defined on an aggregate learning outcome level.

Realism of learning algorithms can be decided on the basis of experimental findings or psychological knowledge (Brenner, 2006). Empirical validity of learning models is usually tested in experimental studies. Erev and Roth (1998) tested their reinforcement learning algorithm against twelve different games with an equilibrium in mixed strategies (nine constant-sum games, three non-constant-sum games); Camerer and Ho (1999) compared simulation results from their EWA model to constant-sum games with unique mixed-strategy equilibria, to a median-action coordination game with multiple Pareto-ranked equilibria, and to a p-beauty contest game with a unique equilibrium. The observation that their algorithms represents human behavior in the studied games in a realistic way, however, does not allow to conclude that they are realistic representations of human learning processes in all (economic) situations. Several further studies compare different learning models to experimental findings for other games or economic contexts (for a comprehensive overview, see Duffy, 2006). Though it would be desirable to have a broad empirical evidence for a large variety of learning algorithms applied in many different economic situations, this is mostly not available. Moreover, outcomes from learning algorithms are usually only compared to laboratory experiments; one might question that human learning and behavior in the artificial stylized setting of those experiments is the same as it would occur "in the field". Given this argument, empirical evidence for computational learning models is even weaker.

It is argued here that this strong empirical evidence is not necessary for all ACE modeling purposes. If agent-based models of every type of game or trading scenario involving learning and adaptation required that the applied learning algorithm is tested in a similar game or scenario against human players, the ACE methodology would loose much of its flexibility. Besides, agent-based research would be restrained to those types of models that can also be examined through human subject experiments. Consequently, the requirement of *realism* of learning representations must be relaxed to some extent. The most important claim for appropriate learning models is that they do not *contradict* psychological knowledge or empirical findings about learning. They should be conform with basic psychological learning principles such as the *Law of Effect* and the *Power Law of Practice* (see Sect. 4.1.1). Ideally, their goodness of fit to empirical learning observation should have been demonstrated in at least one scenario that is relevant to economics.

While these criteria posed on empirical evidence for the learning models applied in AB models are rather weak, it is much more important that simulated learning characteristics are thoroughly assessed in order to filter out parameter values and learning algorithm variants that do not lead to satisfactory learning outcomes. Still better and more robust results can be obtained from agent-based simulations if several suitable learning algorithms are applied to the same scenario under study and their results are compared with each other (i.e. a sensitivity analysis that tests how sensitive results are to the learning representation applied).

4.3 Analysis of Learning Algorithms for Agent-Based Simulations

Under the assumption that all three learning models that have been described in this chapter are, in principle, suitable for agent-based modeling of electricity markets, the question remains how to best set the parameter values and which variants – e.g. which action selection rule or which of the proposed modifications to the Erev and Roth learning algorithm – are best qualified to deliver good simulation results. This assessment is conducted on the basis of four criteria that should be fulfilled by adequate learning model variants.

Criterion 1: Convergence

Do learning results converge to a certain price level?[3] If resulting prices fluctuate as strongly at the end of the simulation as in the beginning, it can be concluded that no learning occurred. Appropriate learning variants and parameter values should lead to a convergence of prices at the end of 2,000 iterations.

Criterion 2: Robustness

How robust are the learning results against varying initial values and random number sequences? Appropriate learning variants and parameter values should show low variability of results under different initial values and random number sequences.

Criterion 3: Collective Minimal Rationality

Do agents succeed to coordinate on bidding sufficient quantities to satisfy total demand? In those cases in which total supply quantity is less than total demand, the market is not cleared and agents receive zero profits. Appropriate learning variants and parameter values should enable agents to jointly bid sufficient quantities.

Criterion 4: Individual Minimal Rationality

Do agents recognize their strategic advantages? Agents with lower marginal generation costs can augment their chances to be a successful bidder through submitting lower-price bids while agents with higher marginal costs have to bid at higher prices in order to avoid losses. Appropriate learning variants and parameter values should lead agents with lower marginal costs to bid at lower prices than those with higher marginal costs.

---

[3] It might also be conceivable that other patterns of learning outcomes, such as cyclic courses of prices, occur. However, apart from outcomes of simulations using the EWA learning model with certain parameter values (as discussed in Sect. 4.3.3), only convergence of prices towards a certain level has been observed.

The four criteria enumerated here are necessary conditions for the suitability of a learning algorithm or variant to adequately model bidding behavior in electricity markets. All learning algorithms or variants that fulfill the four criteria would then have to be tested in a more realistic simulation scenario. If aggregate results from simulations using these selected learning algorithms share the same characteristics as real-world aggregate data, they can be said to be an appropriate representation of agent behavior. This validation procedure is reported in Sect. 5.2.

4.3.2 Simulated Scenario

In order to make the learning algorithms easily comparable, a simplified electricity market is defined. The supply side of this (day-ahead) electricity market consists of five competing agents, each owning a single generating unit. The generating units differ solely in variable generating costs. Their plants' characteristics are summarized in Table 4.1. Variable generating costs are linear in produced electricity quantities, so marginal costs – i.e. the cost of generating the last unit of output – are constant. The demand side is represented by a constant and price insensitive system load. Generator agents learn to set bids that maximize their individual profits.

In contrast to more detailed electricity market simulations as specified in Chap. 5, only one auction takes place per day in this simplified scenario. Thus, the course of prices resulting from the simulation is only based on learning outcomes in a static environment; hourly or seasonal demand variations are not accounted for here, but will be considered in the more detailed settings.

When reinforcement learning is applied, the set of possible actions that an agent can chose from has to be specified *ex-ante*. Possible bid prices have to be limited by a minimum and maximum possible price[4] and discretized over the range between these limits. The definition of the action domain significantly influences simulation results, so it is a crucial task in building agent-based simulations that apply reinforcement learning.

Table 4.1 Agents and power plants in the simplified electricity market scenario

Agent	Plant net capacity (MW)	Variable generating costs (EUR/MWh)
Generator1	500	10
Generator2	500	20
Generator3	500	20
Generator4	500	30
Generator5	500	40

---

[4] After some negative experience with price caps, e.g. at the California power pool, most present day wholesale electricity markets do not impose maximum admissible prices. Upper price limits are at the most fixed for technical reasons. A maximum price in electricity market simulations might, however, be interpreted as a price level at which, if exceeded persistently, regulatory intervention has to be feared, so that it should not occur frequently.

4.3 Analysis of Learning Algorithms for Agent-Based Simulations

In the electricity market scenario developed here, agents submit separate bids for every power plant they own (i.e. one bid in the simplified scenario). A supply bid on the day-ahead market (DAM) consists of a quantity (in MWh) and the price at which the agents wishes to sell this quantity. The agent's action domain spans in discrete steps over the two dimensions price and quantity. Possible prices range from $p^{DAM,min} = 0$ EUR/MWh to $p^{DAM,max} = 100$ EUR/MWh. Possible bid quantities range from 0% to 100% of the plant's available capacity. The price dimension is partitioned into 21, and the quantity dimension, expressing the fraction of available capacity that is bid into the market, is partitioned into six discrete steps. This leads to the following action domain on the day-ahead market:

$$M^{DAM} = [p^{DAM}, \beta^{DAM}] = [\{0,0\}, \{0,0.2\}, ..., \{100, 1.0\}] \quad (4.10)$$

Here, β^{DAM} is the bid fraction of available capacity. The day-ahead market bid b^{DAM} is formulated by multiplying this ratio with the available capacity q^{avail} (which is equal to the net installed capacity in this simplified scenario):

$$b^{DAM} = \left\langle p^{DAM}, (\beta \cdot q^{DAM,avail}) \right\rangle \quad (4.11)$$

In addition to the action domain, the *state domain* which defines all possible states that an agent can be situated in has to be defined for the Q-learning algorithm. In the model developed here, the state domain contains information about the agent's last bid price and about its trading success. As schematized in Fig. 4.2, bid prices can be categorized as low (lower than or equal to one third of the maximum admissible bid price), high (higher than or equal to two thirds of the maximum admissible price) or medium (all remaining prices). A bid can further be categorized as *marginal* or *intra-marginal*, in which case it is a successful bid, or as *extra-marginal* for a bid that was not successful.

The sequence of actions during one simulation iteration is depicted in Fig. 4.3. A trading day starts with the market operator (DayAheadMarketOperator) calling for bids among all registered traders (AdaptiveGenerator agents implementing the interface DayAheadBidder). The agents choose a new action from their reinforcement learning algorithm and formulate a price quantity bid which they submit to the market operator. Market clearing is effected by sorting all supply bids in ascending price order (if two bids have the same bid price, they are sorted in descending volume order; priority between two identical bids is deter-

0	low bid price, (intra-)marginal	1	low bid price, extra-marginal
2	medium bid price, (intra-)marginal	3	medium bid price, extra-marginal
4	high bid price, (intra-)marginal	5	high bid price, extra-marginal

Fig. 4.2 The agent's state domain (with state indices)

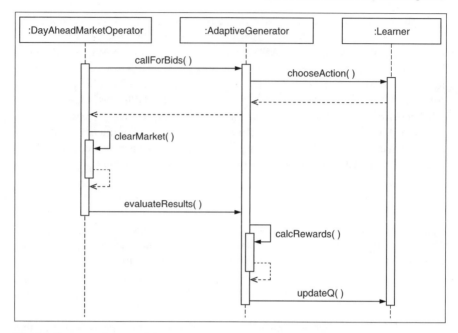

Fig. 4.3 Sequence diagram of the daily trading process on the day-ahead market

mined randomly). The intersection of the so-formed supply curve with the demand curve delivers the price at which all successful bids are remunerated – the market clearing price. The trading results for each bid are sent back to the agents who subsequently evaluate their achieved profits. The reinforcements are calculated on the basis of these profits. In order to prevent negative reinforcements in the case of proportional action selection, the minimum possible profit is subtracted from all profits, as defined by (4.5). The reinforcements are finally determined as the last profit achieved relative to the maximum possible profit; consequently, reinforcements are in the [0..1] range. The propensity, or attraction, or Q-values of the applied learning algorithm is updated by each agent on the basis of this reinforcement. The determination of hypothetical payoffs for the EWA learning model is explicated in Sect. 4.3.3. After the learning update, the next iteration begins.

All learning algorithms and their variants that are described in Sects. 4.1.1, 4.1.2 and 4.1.3 have been tested as behavioral representation of the agents participating in this simplified electricity market. An overview of all tested learning variants and parameter value combinations is given in Table 4.2.

4.3.3 Results

In the following, the average courses of prices and some quantitative measures of learning outcomes for the simplified scenario are described separately for the four criteria of suitable learning variants. One single simulation run constitutes

4.3 Analysis of Learning Algorithms for Agent-Based Simulations

Table 4.2 Tested variants and parameter values of regarded learning algorithms

Learning model	Variant	Determination of action selection probability	Parameter values
Erev and Roth Reinforcement Learning	Original without spillover	Proportional	$\phi = [0.1, 0.2, ..., 0.5]$ $\varepsilon = [0.1, 0.2, ..., 0.5]$ $q_0 = 1.0$
	Original with spillover	Proportional	$\phi = [0.1, 0.2, ..., 0.5]$ $\varepsilon = [0.1, 0.2, ..., 0.5]$ $q_0 = [0.2, 0.4, ..., 1.2]$
	Modified	Proportional	$\phi = [0.1, 0.2, ..., 0.5]$ $\varepsilon = [0.1, 0.2, ..., 0.5]$ $q_0 = [0.2, 0.4, ..., 1.2]$
	Original with spillover	Softmax	$\phi = 0.1$ $\varepsilon = [0.2, 0.3, 0.4]$ $\frac{1}{\tau} = [2, 4, ..., 20]$
	Modified	Softmax	$\phi = 0.1$ $\varepsilon = [0.2, 0.3, 0.4]$ $\frac{1}{\tau} = [2, 4, ..., 20]$
Q-Learning		ε-greedy	$\alpha = [0.4, 0.5, ..., 0.9]$ $\varepsilon = [0.1, 0.2, 0.3]$ $\gamma = [0.8, 0.85, 0.9, 0.95]$ $Q_0 = [0.5, 1.0, 1.5]$
		Softmax	$\alpha = [0.5, 0.6]$ $\gamma = [0.8, 0.85]$ $\frac{1}{\tau} = [2, 4, ...20]$ $Q_0 = [0.5, 1.0, 1.5]$
Experience-Weighted Attraction		Softmax	$\delta = [0.2, 0.3, 0.4, 0.5]$ $\phi = [0.9, 0.95, 1.0]$ $\rho = [0.8, 0.9]$

one possible result from the learning process. As learning contains probabilistic elements, several simulation runs are necessary in order to gain robust results from the simulations and to distinguish outlier outcomes from "normal" runs. Consequently, average results from 50 simulation runs with different random number seeds are presented for all simulations with the Erev and Roth learning model and with Q-learning. Results from the EWA scenario are discussed in a separate section.

Convergence

The analysis of different learning algorithms shows that on average over 50 runs, a convergence of prices to a certain level can be observed with many variants and parameter value combinations of the considered learning models. In contrast, neither of the tested variants exhibits cyclic courses of prices or other patterns or regularities that can be interpreted as the result of collective agent learning (one exception is a special variant of Experience-Weighted Attraction which will be discussed later in this chapter). The fluctuations around the resulting mean price level is stronger for some variants than for others. The level of convergence within one run can be expressed through the variation of market prices during the last 200 iterations of the simulation run. As mean prices may differ for different learning variants, an appropriate quantitative measure for this variance is the *coefficient of variation* (CV), which is defined as the ratio of the standard deviation σ to the mean μ of the last 200 prices: $CV = \frac{\sigma}{\mu}$. Mean coefficients of variance over 50 runs[5] vary between 0.14 and 0.41 for different action selection rules and parameter values of Q-learning. For the original Erev and Roth reinforcement learning, mean CVs vary between 0 and 0.35. Variations are higher for versions without spillover of reinforcement to neighboring actions; the very low variations are only observed for variants with Softmax action selection and very high focus values $\frac{1}{\tau} > 8$. For simulations with the Modified Erev and Roth RL algorithm, prices always converge to one value very quickly with the proportional action selection rule; if Softmax action selection is applied, mean CVs are similar to those observed for the original Erev and Roth algorithm and vary between 0 and 0.33.

Besides the variation of prices, the courses of prices for some learning variants have noticeable slopes at the end of the simulation run. This signals that learning has not stabilized yet and mean price levels would continue to develop if simulations were continued for further iterations. This occurs in particular for those variants of Q-learning with ε-greedy action selection in which the discount factor is high ($\gamma > 0.9$) and α is between 0.4 and 0.7. Based on this result, discount factors above 0.9 for Q-learning are not further considered. Another observed pattern is that the course of prices at the beginning of the simulation is the same as in the end. Such pattern allows the conclusion that the agents have not learned any preferred actions throughout the 2,000 iterations. This occurred only with very low focus parameters $\frac{1}{\tau} < 3$ for the Softmax action selection rule (both with Q-learning and Erev and Roth RL) and for the original Erev and Roth algorithm without spillover and with high experimentation values $\varepsilon \geq 4$. Hence, focus parameter values below 3 are excluded from further examination as well. As will be argued later, Erev and Roth RL without spillover of reinforcement to neighboring actions is no suitable learning model for the simulations scenario studied here.

Examples for courses of prices of average runs for variants in which sensible learning outcomes can be observed for the cases of Q-learning and Erev and Roth RL are shown in Fig. 4.4. Here, the abscissa plots the iterations and the ordinate plots average prices over 50 runs for each iteration.

As a consequence of these first results, the convergence criterion alone is not well suited for distinguishing between appropriate and inappropriate learning variants and parameter values. A further detailed look at the other three quality criteria of learning models is all the more important for the validity of simulation outcomes with adaptive agents.

---

[5] Here, the CV has been calculated for the last 200 iterations of every single simulation run; the means of these CV values over 50 runs are reported.

4.3 Analysis of Learning Algorithms for Agent-Based Simulations

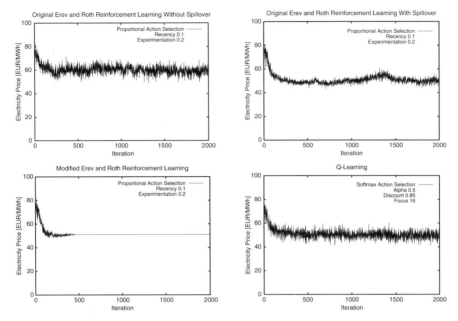

Fig. 4.4 Average course of prices with different learning algorithms (simplified scenario, averaged over 50 simulation runs)

Robustness of the Algorithms

In terms of robustness, it should be assured that outcomes from simulations with one learning algorithm are not too strongly dependent on the chosen initial values for propensities, attractions or Q-values. In addition, it is tested whether the number of available actions and the absolute range of the action domain have an influence on the learning outcome.

As most simulations applying the tested learning variants converge to a certain price level, this level can be taken as the outcome of a simulation run. More precisely, the result of one run is defined as the mean of market prices from the last 200 iterations. For determining the design of experiments of the simulations to be conducted for answering given research questions, it is important to know how strongly simulation results differ from one run to another when different random number sequences are applied. If variations between runs with different random number seeds (RNS) are considerable, than simulations have to be repeated more often, and only mean values over numerous runs can give a reliable quantitative basis for deriving conclusions. Results of the conducted simulations reveal that the variance of results between runs with different RNSs is nearly inversely proportional to the variance of prices at the last iterations within one run (as described in the *convergence* section above). That means that if simulations with one learning variant lead to courses of prices that converge strongly to one price level, the variance of this price level between runs with different random numbers is higher than for those cases in which prices fluctuate stronger around the mean level. The high variance between runs for variants with strong convergence may be an indicator that agents lock in to suboptimal action choices too early and do not continue to explore for better actions. One finding from a comparison of simulation results is that the variance of mean price levels during the last 200 iterations is stronger between runs with the Erev and Roth algorithm. The highest variance is observed for small values of the experimentation parameter ε; simulations with $\varepsilon > 0.1$ vary less strongly and should therefore be preferred. With Q-learning, similar price levels between runs with different random numbers are reached, so this algorithm is more robust against changing random number seeds.

Initial values of propensities or Q-values correspond to an expectation level of the possible outcome for each action. If no pre-game knowledge can be assumed, propensities or Q-values should be equal for all actions. If initial values are very high, than agents have a stronger incentive to try each action until they perceive their real value. A learning algorithm should assure that enough actions are explored, independent of the initial values. The absolute value of initial propensities has to be determined in relation to the order of magnitude that possible reinforcements can have. As the reinforcements chosen here express relative profits (profits achieved divided by maximum possible profits), possible reinforcements range between [0..1]. Initial propensities and Q-values are set to 1.0. A variation of q_0 or Q_0 to 0.5 or 1.5 had no perceivable influence on simulation results, neither for Erev and Roth RL, nor for Q-learning. The considered variants are, thus, robust against varying initial values. Hence, initial Q-values or propensities are always set to 1.0 in further simulations.

In contrast to this, the choice of the action domain, i.e. the definition of all actions that an agent can choose from, has a substantial influence on simulated prices, especially for Q-learning. While the number of discrete actions in a given range is not decisive, its absolute range has a considerable impact on results. Simulations in which the number of prices at which agents can bid their generation output has been reduced from 21 to 11 did not lead to significant differences in simulation outcomes. But when the range of possible prices is extended from originally 0–100 EUR/MWh to 0–200 EUR/MWh, resulting average market prices change considerably. Prices do not simply double – an observation that would correspond to a proportional relationship between bid prices and maximum admissible prices. Insofar, absolute parts, i.e. aspects that are independent of the action domain, play an important role in simulated competition. However, price differences in the case of an altered action domain are significant (see Table 4.3), so the determination of possible actions that an agent can choose from has to be conducted with great care.

Similar results are produced when the possible bid price range is lowered to 0–60 EUR/MWh. A comparison of the courses of prices for three different action domains with maximum admissible bid prices of 60, 100, and 200 EUR/MWh is depicted in Fig. 4.5 for the Erev and Roth reinforcement learning algorithm.

4.3 Analysis of Learning Algorithms for Agent-Based Simulations

Table 4.3 Average market prices of last 200 iterations for simulations with varied action domains

Learning algorithm (one respective variant)	Bid price range 0–100 (EUR/MWh)	Bid price range 0–200 (EUR/MWh)
Original Erev and Roth RL[a]	49.92	70.41
Modified Erev and Roth RL[b]	51.10	70.6
Q-Learning[c]	49.69	83.75

[a]With spillover of reinforcements, proportional action selection, $\phi = 0.1$, $\varepsilon = 0.2$, $q_0 = 1.0$
[b]Proportional action selection, $\phi = 0.1$, $\varepsilon = 0.2$, $q_0 = 1.0$
[c]Softmax action selection, $\alpha = 0.5$, $\gamma = 0.85$, $\frac{1}{\tau} = 16$

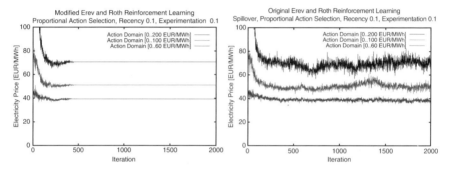

Fig. 4.5 Average course of prices with Erev and Roth learning for different action domains

Collective Minimal Rationality

The criterion whether agents bid profit maximizing supply quantities is a further indicator of learning variants that are more or less appropriate for an agent's behavioral representation. It can be observed that both the original and the modified formulation of the Erev and Roth algorithm with proportional action selection often lead to bid rounds in which the total supply quantity is not sufficient to satisfy demand. The frequency of this insufficient demand situations is increasing in both recency and experimentation parameter values. It becomes very frequent when the recency parameter rises above $\rho > 0.2$, so the recency parameter should be set to small values (0.1 or 0.2). With Softmax action selection, all Erev and Roth variants display many rounds with insufficient supply bid quantity for low and for very high focus parameters; for values of $6 \leq \frac{1}{\tau} \leq 12$, there are hardly any rounds in which this situation occurs, so these parameter values are more appropriate for further simulations. Q-learning leads to few bid rounds with insufficient supply quantity at the end of a simulation run with all variants.

Individual Minimal Rationality

A view on resulting bid behavior under application of different learning algorithms reveals that also under those variants that exhibited satisfying characteristics under the preceding three criteria, there are still some variants in which agents do not recognize their strategic advantages. In those cases either all agents bid at similar prices or there are cases in which agents with lower marginal costs bid at higher prices than those with higher marginal costs, and vice versa. It can be assumed that in more complex scenarios, these variants are no longer appropriate for realistically representing agent learning. Consequently, these variants or parameter combinations are not considered for further simulations.

Among those variants that have not been rejected on the basis of the first three quality criteria, the following lead to simulation results in which either all agents bid at the same prices on average, or in which the order of agents sorted by bid prices is different from the order when they are sorted by marginal costs: MRE with $\phi = 0.2$ and $\varepsilon > 0.1$, Erev and Roth variants with Softmax action selection and $\frac{1}{\tau} > 8$, Q-learning with ε-greedy action selection and low discount rates in combination with low exploration rates ($\gamma = 0.8$, $\varepsilon = 0.1$) or with very high learning rates ($\alpha > 0.9$). All these variants are not suitable for representing learning in more complex scenarios.

Figure 4.6 shows average bid prices of individual agents as a function of their marginal costs. The curves, respectively, represent the same variants as those depicted in Figs. 4.4 and 4.5 and in Table 4.3, i.e. variants that are deemed appropriate for representing agent learning in electricity market simulations of the kind considered here. The figure highlights that agents learn to recognize their strategic advantages and to bid accordingly when applying appropriate learning algorithms.

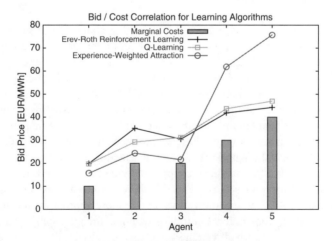

Fig. 4.6 Average agents' bid prices as a function of marginal generation costs

This last quality criterion for learning algorithms is the most insightful one. Learning variants that do not exhibit satisfying results on the basis of the first three criteria also perform poorly under this fourth criterion. Conversely, among those performing seemingly well when considering the first criteria, there remain a few that can not be considered suitable for representing agent learning behavior when analyzed under the individual rationality criterion. This underlines the importance of a thorough quality analysis of variants and parameter combinations for candidate learning models.

Evaluation of Simulations with Experience-Weighted Attraction

Simulations applying the Experience-Weighted Attraction algorithm show more interesting result patterns. With this learning model, it has to be decided how hypothetical rewards are determined for those actions that have not been chosen. One obvious naïve approach is to assume that an agent is a price-taker, so his bid does not influence resulting market prices. Under this assumption, hypothetical rewards would be equal to the market clearing price multiplied by the bid quantity and reduced by generating costs for all bids below the market clearing price. For bids above the market clearing price, the resulting profit would be equal to zero. If the bid price submitted equals the resulting market clearing price, it might be assumed that this bid would have been successful and the resulting profit could be calculated accordingly. Simulations with this approach for calculating hypothetical rewards lead to market prices that equal the system marginal cost (the cost of the last unit necessary to satisfy demand) after only a few iterations, and stay at this level throughout the rest of the simulation. This is no realistic outcome for oligopolistic competition.

The assumption that single generators cannot influence prices is only adequate in a perfectly competitive market. In an oligopoly like the market simulated here, every agent has a certain probability to be the price-setter. This raises the question how high the probability is that a bidder would have had been successful if he had bid at a price above the market clearing price. An agent in the simulation scenario presented here (and in real-world markets where bid curves are not published) is not able to answer this question on the basis of the information he has about the market. He can only have an estimate of this probability, on the basis of which he can calculate his hypothetical profits. In this case, hypothetical profits are uncertain, and different agents might have different estimates of the probability to be successful with a higher bid price.

In further simulations, the described probability has been assumed to be 10% for all agents. The hypothetical reward for an action that implies a bid prices above the market clearing price is determined by the product of the profit that would be gained if the bid was successful, multiplied with the success probability. Resulting courses of prices for this setting are depicted in Fig. 4.7 (left).

It can be seen that also for this variant of the Experience-Weighted Attraction algorithm, simulated prices are significantly lower than those from simulations with

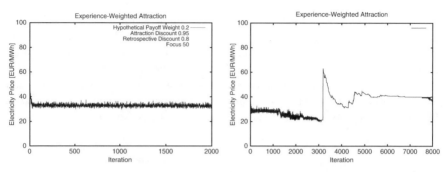

Fig. 4.7 Average course of prices with Experience-Weighted Attraction (simplified scenario)

Erev and Roth reinforcement learning or Q-learning. However, prices are now above system marginal costs. Simulations with an assumed probability of success for bid prices above the market clearing price of 20% deliver a similar picture.

Extraordinary courses of prices can be observed if the attraction discount factor ϕ is set to a value of 1.0 (see Fig. 4.7, right). In this case, market prices decrease steadily at the beginning of the simulation, until they reach the level of system marginal costs. Subsequently, prices show an abrupt increase, and then fall back relatively quickly until they stabilize at a higher price level. This can be explained by the fact that the attractions of actions with high bid prices (and a high risk of not being successful) adopt values that lie above those actions with low bid prices (and high success probability) after a certain number of iterations. High bid prices are then chosen by several agents, leading to high market clearing prices. As a consequence, attractions of all actions with bids below these high market clearing prices rise quickly, so prices continue to be high for some time and do not immediately jump back to their previous value after a price peak.

The analysis of the agents' bidding behavior in simulations applying Experience-Weighted Attraction shows another interesting characteristic. It can be observed that average bids resemble the *hockey-stick bidding* phenomenon. This characterizes a supply curve that looks similar to a hockey-stick: low, and only slowly increasing prices for the largest part of the supply quantity, and a sharp increase in prices for the last quantities bid into the market. Such a supply curve results from different strategies for different generating units: most bidders submit low bids in order to increase their probability to be successful. Bidders with very high marginal costs assume that the probability of a marginal cost bid to be within the set of successful bids is low. In case they are the price-setting seller with a marginal cost bid, their profit would be zero, just like in the case when their bids are not successful. An even higher bid would not significantly lower the probability of being successful and would not lead to lost profits. But for those cases in which the demand and supply situation is such that they are called into operation, bidders with high marginal costs want to make a high profit. Consequently, their strategy is to bid at high prices. Mount (2000) ascribes the appearance of very high price peaks in the electricity market of Pennsylvania, New Jersey and Maryland (PJM) to such a bidding behavior.

As in the research work at hand the interplay between different market is in the center of interest, a learning model should be as generic as possible and applicable to many bidding situations. The introduction of the agent's perceived probability of being successful with a bid price above the market clearing price introduces an additional parameter which must be carefully analyzed before it is applied to more detailed simulation scenarios. This analysis and discussion cannot be delivered here. In order to avoid to further complicate the model or to make it less easily comprehensible and traceable, it will be restrained from further considering Experience-Weighted Attraction for the agent-based simulation of interrelated electricity and emissions markets. However, the EWA model is an interesting alternative to reinforcement learning; the analysis of its appropriateness for other research questions dealing with electricity markets is an interesting question for future research.

4.3.4 Implications for Robust and Valid Agent-Based Simulations

On the basis of simulation outcomes from the simplified scenario, those variants and parameter value combinations that will be applied in further simulation runs are selected. Simulations applying the original Erev and Roth reinforcement learning algorithm as formulated for unsorted action domains performed poorly. For all tested parameter values, the variants which included a spillover of reinforcement to the neighboring actions performed significantly better. This latter variant has been formulated for action domains in which similarity of actions can be linearly ordered, which is the case in the scenario presented here. Thus, all variants without a spillover of reinforcements as defined by (4.2) (with the modifications defined in Appendix A.1) are excluded from further considerations.

An overview of the selected learning variants is given in Table 4.4. More detailed characteristics of the selected variants in terms of the criteria analyzed in this chapter are provided in Appendix A.2. It becomes obvious that aggregate simulation outcomes are quite similar with Erev and Roth reinforcement learning and with Q-learning. This finding strengthens confidence in the appropriateness of these learning models to realistically represent strategic agent behavior in electricity markets.

It has been argued that evolutionary concepts should only be applied if more simple models of learning fail to deliver realistic simulation outcomes. After an analysis of the reinforcement and belief-based learning algorithms presented here, it can be concluded that these are appropriate for representing adaptive agent behavior in an electricity market scenario in which price-quantity supply bids have to be optimized. Based on this finding, it is judged not necessary to test further learning concepts for the problem analyzed in this work.

A diligent analysis of the behavioral (adaptive) agent representation is a necessary – notwithstanding not sufficient – condition for valid agent-based simulations. After being qualified as suitable learning models in the simplified scenario, the variants have to be applied in a more realistic scenario as a next validation step. There,

Table 4.4 Appropriate variants of learning algorithms

Learning model	Variant	Parameter values
Erev and Roth Reinforcement Learning	Original with spillover, proportional	$\phi = [0.1, 0.2]$ $\varepsilon = [0.2, 0.3, 0.4]$
	Modified, proportional	$\phi = [0.1, 0.2]$ $\varepsilon = [0.2, 0.3, 0.4]$
	Original with spillover, Softmax	$\phi = 0.1$ $\varepsilon = [0.2, 0.3, 0.4]$ $\frac{1}{\tau} = [6, 7, 8]$
	Modified, Softmax	$\phi = 0.1$ $\varepsilon = [0.2, 0.3, 0.4]$ $\frac{1}{\tau} = [6, 7, 8]$
Q-Learning	ε-greedy	$\alpha = [0.4, 0.5, ..., 0.8]$ $\varepsilon = [0.1, 0.2, 0.3]$ $\gamma = [0.8, 0.85, 0.9]$
	Softmax	$\alpha = [0.5, 0.6]$ $\gamma = [0.8, 0.85, 0.9]$ $\frac{1}{\tau} = [4, ...20]$

it will become obvious whether an agent-based simulation model with behavior represented by the selected learning variants is able to produce empirically observed stylized facts (in Sect. 5.2 this inspection is described for the model developed in this work).

As soon as an agent-based simulation model is designed to represent a realistic scenario that is able to reproduce empirical data, it has to be fitted – or calibrated – to empirical (historical) data. Besides the learning parameters, there are usually numerous further variables of the simulation model to which parameter values have to be attributed. If all parameters of an agent-based models are jointly calibrated, there is a high risk of *overfitting* the model. In such a case the final fitted model parameters would be arbitrary and unable to generalize beyond the fitting data. Through analyzing the impact of different learning parameters separately on the basis of clearly defined, traceable quality criteria and for a simplified scenario which resembles the realistic scenario, the degrees of freedom for the more complicated model are reduced and the risk of overfitting decreases. One variable that remains to be calibrated with the realistic model is the range of bid prices that agents can choose from, constituting the action domain.

The procedure proposed in this chapter is a useful way of *micro-validating* the learning part of the agent-based simulation model. Similar procedures should always be carried out when building a model that contains adaptive agents.

4.4 Summary

In this chapter, different learning models that have been applied in the economics literature were presented. The learning algorithms are either motivated by psychological findings on human learning processes or stem from the computer science field of *machine learning*; all algorithms are possible means of modeling strategic agent bidding behavior in electricity markets. The question to be answered in this chapter was which learning model is most appropriate for the behavioral representation of agents that compete for selling electricity, thereby seeking to optimize their bids which consist of price quantity pairs.

For a thorough assessment of the algorithms, a simplified electricity market scenario has been defined. Three learning models have been tested for appropriateness in this simplified model: Q-learning as formulated by Watkins (1989), the learning algorithm formulated by Erev and Roth (1998) as well as a variation of this algorithm proposed by Nicolaisen et al. (2001), and the Experience-Weighted Attraction model formulated by Camerer and Ho (1999). For each of these models, numerous different parameter values have been considered.

As a result, it can be stated that reinforcement learning algorithms such as Q-learning and the Erev and Roth learning model are well suitable for representing agent behavior in electricity trading scenarios such as the simplified market model presented here. Experience-Weighted Attraction on the other hand only delivers realistic results when special assumptions about the agents' reasoning are made. As the reinforcement learning algorithms already deliver satisfactory results, the EWA model will not be considered for further simulations.

The suitable variants and parameter value combinations of Q-learning and Erev and Roth reinforcement learning that come out of the analysis of this chapter will next be applied in the more detailed simulation scenarios described in the following chapters.

Chapter 5
The Electricity Sector Simulation Model

In this chapter, a realistic model of the German wholesale power and emissions trading markets is described. The adaptive agents that represent electricity generators in this model implement the reinforcement learning algorithms that have performed best in the analysis conducted in the preceding chapter.

In Sect. 5.1, the overall structure of the model and the agent behavior are presented in detail. In a next step, it is tested whether the model applying the chosen learning algorithms is able to reproduce stylized facts observed at real-world electricity and CO_2 allowance markets in Germany (Sect. 5.2.1). A sensitivity analysis (Sect. 5.2.2) gives further information about the most important factors influencing simulation results. With the validated model, simulation runs with well defined market design scenarios are then carried out. Results from these simulations help to answer the research questions treated in this work; these are presented in Chap. 6.

5.1 Design of the Simulation Model

An overview of the simulation model architecture is provided in Sect. 5.1.1. The model comprises three markets, which are described subsequently: the day-ahead electricity market (see Sect. 5.1.2), the market for balancing power (see Sect. 5.1.3), and a market for CO_2 emission allowance trading (see Sect. 5.1.4). The markets are interrelated through the agents' bidding strategies; these aspects are described in Sect. 5.1.5. In order to make the simulation model accessible for decision makers in the electricity sector, a graphical user interface (GUI) has been implemented that helps to easily define various simulation scenarios. This GUI is presented in Sect. 5.1.6, along with further implementation details that characterize the developed simulation model.

5.1.1 Overall Model Structure

Following the agent-based paradigm, all relevant parts of the electricity sector simulation model are modeled as *agents*. The set of agents comprises the market operators, power plant operators who generate electricity and sell their generation output either on the day-ahead electricity market or on the balancing power market and who buy or sell CO_2 emission allowances, load serving entities who demand electricity on the day-ahead market, and CO_2 market participants who constitute the external demand and supply of allowances. All agents inherit basic methods from the abstract `PowerACEAgent` class. The markets are subclasses of the `MarketOperator` agent class, and market participants inherit parts of the `Trader` agent class. In order to ensure that agents can submit bids of the correct `Bid` subclass, they have to implement the corresponding interface of the market, i.e. `DayAheadBidder`, `BalancingBidder`, or `CO2Bidder`. The simulated agents and their inheritance structure are graphically depicted in Fig. 5.1.

In the simulations presented in this chapter, agents of the class `Adaptive Generator` are able to trade on all three markets (the class implements all three market interfaces), and they can act strategically on the two power markets (through the interface `AdaptiveAgent`). Agents of the type `LoadServingEntity` do not act strategically; however, a new class `AdaptiveLoadServingEntity`, which corresponds to an actively bidding demand side agent on the day-ahead electricity market, can easily be defined; it must then implement the `AdaptiveAgent` interface and instantiate a learning algorithm class.

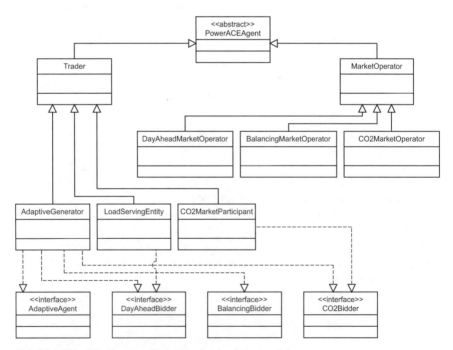

Fig. 5.1 UML class diagram of agents in the simulation model

5.1 Design of the Simulation Model

The model implementation uses the *Recursive Porus Agent Simulation Toolkit* Repast, which is a JAVA-based class library that facilitates agent-based simulations (for a discussion of software tools supporting agent-based simulations, see Sect. 3.2.3). Repast offers a GUI with the help of which basic functions of the model can be controlled, such as starting and stopping a simulation run, graphical outputs, etc. The methods corresponding to these actions have to be specified by the modeler; the `buildModel()` method, for instance, specifies all actions that have to be effected when starting a new simulation run, and `step()` defines the actions to be taken at every simulation iteration.

In the model developed here, a limited number of agents repeatedly submit bids to several markets. One iteration within the simulation corresponds to one trading day. Generators who operate fossil fuel fired power plants can offer their capacity on the power markets and also trade CO_2 emission allowances at the carbon exchange. Those generators that own power plants which meet the technical requirements for delivering tertiary balancing power (minute reserve) can also bid their capacity into the balancing power market. Generator agents seek to maximize their individual profits and have private information about their generating costs and capacities.

As depicted in Fig. 5.2, at the beginning of a new simulation run all agents are instantiated and initialized (here, only initialization of market operators is shown; trader agents also initialize their reinforcement learning algorithms at the

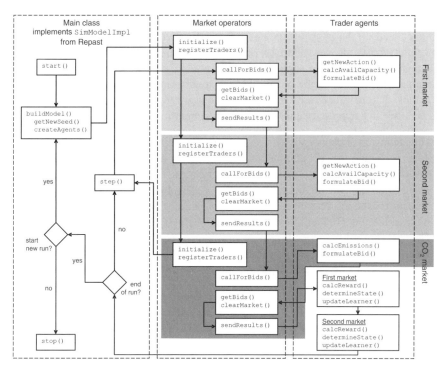

Fig. 5.2 Flowchart of the actions performed at each simulation iteration

beginning of a new run). Every trading day starts with the first market operator calling for bids among all registered traders. The order of market clearing can be varied, i.e. either the day-ahead electricity market or the balancing power market can be cleared first, or both can be cleared in the same hour. All agents implementing the interface `DayAheadBidder` can respond to the call for bids by the `DayAheadMarketOperator`, and those who implement the `Balancing Bidder` interface can respond to the call by the `BalancingMarketOperator`.

For each market they participate in, agents choose new actions from their reinforcement learning algorithms and formulate according bids which they submit to the market operator (see Sects. 5.1.2 and 5.1.3 for a description of the bid formulation). Agents keep reference to the list of bids they submit to the market. During the clearing process, the market operators write trading results into the bid instances, i.e. the resulting volumes and the market price or, in case of discriminatory pricing, the individual matching price of each bid. Through this procedure, agents can later iterate over their bid list and determine the rewards they gained from trading, in order to feed back the results to their learning algorithm instances.

After both power markets have been cleared, agents calculate the CO_2 emissions associated to their sold electricity output. In case they need additional allowances to cover their emissions, agents formulate demand bids; if they have a surplus of allowances, they can submit offer bids. The CO_2 bid procedure is described in more detail in Sect. 5.1.4.

At the end of an iteration, when all three markets are cleared, the adaptive agents update their reinforcement learners. The reinforcements and, if applicable, the states are calculated on the basis of market prices and individual trading success, taking opportunity costs into account (see Sect. 5.1.5 for further explanation of the reinforcements and considered opportunity costs).

Supply side agents on the power markets are characterized by their power plant portfolios. The parameters of each power plant are the following (see Tables B.2 and B.3 for the parameter values assumed for simulations):

- The net installed capacity
- (Constant) marginal generation costs
- No-load costs
- The availability (this parameter is used for renewable energy sources in order to account for the discontinuous availability of these sources)
- A classification whether the plant is technically able to deliver minute reserve
- The emission factor, denoting how much CO_2 emissions are associated with every MWh of output
- The initial allocation of CO_2 emission allowances granted for the plant

Demand side agents on the day-ahead electricity market are characterized by their (hourly or daily) load in MWh. Emissions trading participants other than the generator agents are characterized by their demand or supply quantity and a valuation at which they wish to sell or buy CO_2 emission allowances.

The electricity sector simulation model does not take into account any transmission constraints. The German electricity sector disposes of a tightly intermeshed

5.1 Design of the Simulation Model

transmission system with large transportation capacities, so electricity transportation within Germany is almost always possible. However, future developments in the power industry, such as an increased deployment of wind energy sources in regions with low demand, may lead to situations in which transmission constraints are binding and regional prices have to be calculated (dena, 2005). The inclusion of the transmission grid is a possible enhancement of the model and may be subject to future work.

5.1.2 The Day-Ahead Electricity Market Model

Two different realizations of the day-ahead wholesale power market have been implemented. One implementation has been designed in accordance with the rules in place at the daily *spot* market auction operated by the European Energy Exchange AG (EEX, 2007, see Sect. 2.1.1 for a brief summary). Several block contracts, as well as hourly contracts can be traded, and block contracts are taken into account in the calculation of hourly prices through an iterative process.

For the research questions at hand, however, it was judged sufficient to model the day-ahead market as a sequence of 24 simple call markets for every delivery hour of the following day. Agents submit bids for hourly contracts; block bids are neglected in the simulations described here. The specifications of the hourly call markets, in which electricity delivery contracts are traded, correspond to the market defined in the simplified electricity trading scenario, which is described in Sect. 4.3.2.

Each agent i owns a set of G_i generating units (power plants). For each generating unit, the available capacity that an agent can offer in the day-ahead market depends on the trading results on the balancing power market (BPM), if the latter is cleared earlier (which is assumed in the following simulation scenarios). If an agent has sold (part of) a plant's capacity on the balancing power market, his available capacity q^{avail} is no longer the total net installed capacity q^{net}, but is reduced by the amount of sold (or committed)[1] capacity of this plant, q^{comm}. The available capacity is calculated separately for every hour h; the corresponding committed quantity is the sold or bid capacity of the bidding block that contains the specific hour, $k(h)$:

$$q^{DAM,avail}_{h,i,g} = q^{net}_{i,g} - q^{BPM,comm}_{k(h),i,g} \tag{5.1}$$

On every trading day, an agent can submit a set of hourly bids for each power plant it owns. One supply bid consists of a bid quantity and the price at which the quantity is offered. The agent formulates the bids according to the output of the

---

[1] It is also possible to define trading scenarios in which results from the first market are not yet published when the second market calls for bids. In this case, an agent who has bid capacity on the first market cannot bid this quantity again in the second market, because he would face the risk of not having enough capacity to fulfill his traded contracts if both bids are successful. Generators are modeled here as suppliers only; they cannot buy electricity for reselling purposes. Consequently, q^{comm} here corresponds to the bid quantity on the first market.

reinforcement learning algorithm, where the learning domain is defined by (4.10) (see preceding chapter). The set of bids submitted to the day-ahead market by agent i for delivery hour h, which contains separate bids for each generating unit g ($b_{h,i,g}^{DAM}$) is denoted $B_{h,i}^{DAM}$ and is defined by (5.2) (in analogy to (4.10) of the simplified scenario, $\beta = 0, 20, ..., 100\%$ is the fraction of available capacity of plant g that agent i bids into the market for delivery in hour h):

$$B_{h,i}^{DAM} = \{\langle p_{h,i,g}^{DAM}, q_{h,i,g}^{DAM} \rangle : g = 1, ..., G_i\} \quad (5.2)$$

with $q_{h,i,g}^{DAM} = \beta_{h,i,g} \cdot q_{h,i,g}^{DAM,avail}$ with $0 \leq q_{h,i,g}^{DAM} \leq q_{h,i,g}^{DAM,avail}$

The set of actions defined in the action domain ensures that the capacity constraint condition $0 \leq q_{h,i,g}^{DAM} \leq q_{h,i,g}^{DAM,avail}$ is always met. If β_h is equal to zero, no bid is formulated for this hour, as bid volumes have to be higher than zero. This situation can be interpreted as total capacity withholding for the respective plant.

Market clearing is effected by sorting all supply bids in ascending, and demand bids in descending price order. In case of equal bid prices, bids are sorted in descending volume order; priority between two identical bids is determined randomly.[2] The intersection of the so-formed supply and demand curves sets the resulting price P_h^{DAM} at which all successful bids are remunerated.

The demand side of the day-ahead market is represented as a fixed price-insensitive load.[3] Data of the hourly system's total load is used for representing electricity demand. The assumption of a fixed load is a useful simplification. In the short-term, it is also realistic to assume that electricity consumers do not react to price changes, because they usually do not have any price information at short notice that allows them to adapt their consumption to the price signals. As the research questions investigated here focus on short-term market dynamics, fixed price-insensitive load is a valid assumption. As it was the case in the real-world system during the time frame under study (2006), overall electric load L never exceeds the total installed net capacity of all generators, i.e.

$$L_h \leq \sum_i \sum_{g=1}^{G_i} q_{h,i,g}^{net} \; \forall h \quad (5.3)$$

The last successful bid in this *merit order dispatch* procedure sets the price; it is denoted as the *marginal* bid. In this equilibrium point, the sum of all supply quantities equals the total demand quantity. All capacities that were bid at prices below the market price P_h^{DAM} (these are called *intra-marginal* bids) are dispatched with their full bid quantity. The dispatch quantity for the marginal bid is the remaining

---

[2] Before the clearing process, all bids are randomized in order to avoid unintended effects in cases of identical bids.

[3] The model implementation also allows to represent demand as strategic agents who bid actively and learn to set the prices allowing them to maximize the margins they obtain from reselling the electricity to their customers at the retail level. This model feature might be helpful for specific research questions, but is not applied in this work.

5.1 Design of the Simulation Model

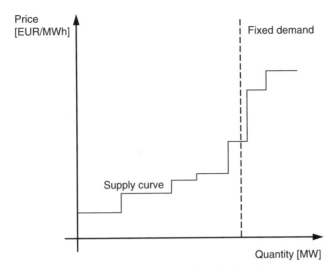

Fig. 5.3 Illustrative supply and demand curves at the day-ahead market

demand, and all bids at prices higher than the market price (*extra-marginal* bids) are not dispatched.[4]

For illustration of the market clearing, a schematic example for an equilibrium of a bid supply curve with a fixed price-insensitive demand is depicted in Fig. 5.3. If curves intersect on a vertical line segment, i.e. infinitely many prices along this segment are possible, the lower end of the segment is defined as the clearing price.

If the sum of all supply bid quantities is less than total demand, two proceedings are possible. The *spot market concept* employed at the European Energy Exchange (EEX, 2007) defines a *proportionate execution* for cases in which no intersection of supply and demand curves can be established. This procedure is also implemented in the model, but in the simulations presented here, a second, simplified variant of dealing with such market rounds is used: in those hours in which supply quantity is less than demand, no market clearing is carried out at all. The resulting price in this case is set to the maximum possible price of the market, and no volumes are traded.

For some types of power plants, the possible actions that an agent can take differ from (4.10). Nuclear power plants and lignite-fired power plants do not allow short-term load changes, but have to be kept at a relatively constant or slow-changing power rating. Therefore, it is not realistic to assume that these power plants are deployed for strategic bidding of hourly power delivery. It is assumed here that these power plant types bid their whole capacity into the day-ahead market at their respective marginal costs.

Furthermore, it is assumed that weather forecasts are not yet precise enough for predicting the output power of wind energy converters in every hour of the following

---

[4] If there are several bids at a price equal to P_h^{DAM}, the rank of the bid within the merit order, which is higher for higher bid volumes and random for equal bids, determines the dispatch quantity.

day. Consequently, electricity from wind energy can not be bid strategically at the day-ahead market. In order to take into account the electricity amount produced by wind turbines, the installed wind energy capacity of the reference scenario year 2006 is multiplied with yearly average full load hours (*availability*) for estimating the capacity that is available in every hour. This quantity is bid into the day-ahead market at a price equal to marginal generating costs for the respective units.

The problem facing the agents who bid on the day-ahead market and the reinforcements they use as feedback from the market are described in Sect. 5.1.5. The data input for simulation runs representing the German electricity market are declared in Sect. 5.2.1.

5.1.3 The Balancing Power Market Model

The implemented balancing power market represents procurement auctions for positive minute reserve, and is designed in a similar fashion as the minute reserve auctions operated by the four German transmission system operators (see Sect. 2.1.2 for a description). More precisely, the specifications defined in a recent adjudication issued by the responsible regulatory authority (Bundesnetzagentur, 2006) are already represented in the BPM implemented in the electricity sector simulation model.

Agents with power plants that are technically able to deliver minute reserve (MR) can sell capacity on this market. These plants have to allow fast changes in load and must be ready to be fully activated within 15 min. In the simulation model developed here, only gas-fired and hydroelectric power plants are assumed to be capable of delivering minute reserve. The capability of minute reserve delivery is expressed through the indicator function $I(g, \text{MR})$ which is equal to zero if plant g is slowly controllable (not suitable for minute reserve delivery) and one if it is fast controllable and can, thus, bid on the BPM.

A bid on the balancing power market contains an offer quantity in MW and two bid prices: the *capacity price* is the price for holding capacity in reserve during the whole bidding period; the *energy price* is the price a generator is paid for produced minute reserve in case his plant is actually deployed for regulating purposes.

Similarly to the day-ahead market, the domain of possible actions on the market for balancing power market is two-dimensional; the two dimensions are capacity prices and energy prices. Possible prices range from 0 to 200 EUR/MW in 21 discrete steps for the capacity (*cap*) price and from 0 to 100 EUR/MWh in five steps for the *energy* price. This leads to the following action domain:

$$M^{BPM} = \left[p^{BPM,cap}, p^{BPM,energy}\right] = [\{0,0\}, \{0,25\}, ..., \{200,100\}] \quad (5.4)$$

If the balancing power market is cleared after the day-ahead market, the agent first calculates the available capacity of all plants that can deliver minute reserve. As each bidding block k contains several hours, the maximum committed capacity of all hours that are part of the bidding block, $h \in k$, constitutes the binding constraint for

5.1 Design of the Simulation Model

the remaining available capacity. Agents always bid their whole available capacity into the balancing power market.

$$q_{k,i,g}^{BPM,avail} = q_{i,g}^{net} - \max_{h \in k} q_{h,i,g}^{DAM,comm} \quad (5.5)$$

After deciding which action to take, the agent formulates the bid that it wants to submit to the `BalancingMarketOperator`; the specification of a balancing power bid is given in (5.6).

$$B_{k,i}^{BPM} = \left\{ \left\langle p_{k,i,g}^{BPM,cap}, p_{k,i,g}^{BPM,energy}, q_{k,i,g}^{BPM} \right\rangle : g = 1, \ldots, G_i | I(g, \text{MR}) = 1 \right\} \quad (5.6)$$

$$\text{with } q_{k,i,g}^{BPM} = q_{k,i,g}^{BPM,avail}$$

If Q-learning is applied in the simulation, different *states* also have to be defined for the balancing power market. These are the same as in the day-ahead market, and are described in Sect. 4.3.2. In the model, only capacity prices are considered for state determination; capacity bid prices are categorized as low (lower than or equal to one third of the maximum admissible bid price), high (higher than or equal to two thirds of the maximum admissible price) or medium (all remaining prices). A bid is further categorized as marginal or intra-marginal, in which case it is a successful bid, or as extra-marginal for a bid that was not successful. The resulting six states are schematized in Fig. 4.2 (see preceding chapter).

The demand side of the balancing power market is represented as a predefined quantity of positive minute reserve that has to be procured. Six equally long bidding blocks of 4 h length are differentiated for every trading day: from 0 am to 4 am, from 4 am to 8 am, and so forth. The tendered balancing capacity quantity Q_k^{BPM} is equal for every bidding block (here, $Q_k^{BPM} = 3{,}500 \; \forall k$).

In procurement auctions with two-part bids, the market clearing procedure distinguishes two aspects: the *scoring rule* and the *settlement rule* (Chao & Wilson, 2002). The scoring rule defines how to compare bids and the settlement rule determines payments. In the balancing market developed here, bids are compared solely on the basis of capacity bid prices, i.e. the bids with the lowest capacity bid price are considered first for holding capacity in reserve, until demand is met. If minute reserve is actually needed for frequency control on the day of delivery, deployment is decided among those bids that had been successful on the corresponding trading day. A merit order is constructed based on the energy prices only, and bids with lowest energy prices are deployed first, until demand is met. Payments both for holding capacity in reserve and for delivering energy for frequency regulating purposes are based on the respective bid prices. These scoring and settlement rules both correspond to the specifications set by Bundesnetzagentur (2006); a variation of the settlement rule is discussed in Sect. 6.2.[5]

---

[5] In her adjudication, the Bundesnetzagentur (2006) states the problem that a scoring rule that is solely based on capacity prices offers some possibilities for the market participants to game the

5.1.4 The Model of Emissions Trading

The CO_2 emission allowance market (CO2M) is modeled as a sealed bid double-auction that is cleared at the end of each trading day. Supply and demand bids are summed up to form supply and demand functions in the same way as on the day-ahead electricity market, and the uniform market clearing price is determined by the intersection of both curves.

Bids on the CO_2 market contain a volume of allowances[6] that is offered or asked, a bid price, and the compliance period (cp) for which the allowance should be valid (for simplicity, and because no speculation is considered in this model, the compliance period is always set equal to the current period). Buying bids have positive volumes, and selling offers have negative volumes. Each agent submits one single bid on the allowance market at time t (day of the year y), representing its daily allowance requirement or surplus which is calculated for the whole portfolio of power plants it owns.

$$b_{t,i}^{CO2M} = \langle p_{t,i}^{CO2M}, q_{t,i}^{CO2M}, cp \rangle \tag{5.7}$$

All generator agents that own fossil fuel fired power plants are initially endowed with a certain amount of CO_2 allowances. The initial allocation of allowances is calculated according to a grandfathering rule, i.e. based on past emissions for each single power plant. The sectors outside the electricity industry that are covered by the emissions trading scheme submit a fixed supply and demand every day. As little is known about CO_2 mitigation costs of these sectors – and consequently about their valuation for certificates – their supply and demand is calibrated so that average prices that arise endogenously during the simulation roughly correspond to observed prices in the real-world carbon exchanges. The characteristics of all market participants are summarized in Appendix B.3, (for power generating agents) and B.4 (for the other industries covered by the EU-ETS).

It is assumed that all agents seek to even up their open positions every day. This entails that agents who sell electricity also make sure to have enough allowances for the carbon dioxide emissions associated to their generation output. Agents who have surplus allowances try to sell them at the market price. Market participants in real-world carbon exchanges also have the possibility to buy (sell) more allowances than currently needed if they expect prices to rise (fall) in the future. Strategies of this kind, i.e. buying and selling allowances to profit from price changes instead of using them directly for compliance, can be categorized as speculation. Speculation is not considered in this model.

The agents' daily trading quantities are calculated on the basis of initial endowments $q_y^{CO_2,init}$ issued for year y, and of trading success on the current trading day. The amount of carbon dioxide emitted during electricity generation, $q^{CO_2,emit}$, is determined by the electricity amounts sold at the day-ahead market $q^{DAM,sold}$ and by

---

mechanism. This conclusion is followed by the request to conduct further research on incentive compatible market rules for balancing power markets.

[6] 1 EUA (= EU-Allowance) allows for the emission of 1 ton of CO_2.

5.1 Design of the Simulation Model

deployed minute reserve $q^{BPM,depl}$.[7] The quantities are multiplied with the emission factor ω_g of plant g, quantifying the CO_2 emissions associated with every MWh of power output generated from that plant.

$$q_{t,i}^{CO_2,emit} = \sum_{G_i} \left(\sum_{h=1}^{24} \omega_g \cdot q_{h,i,g}^{DAM,sold} + \sum_{k=1}^{6} \omega_g \cdot q_{k,i,g}^{BPM,depl} \right) \tag{5.8}$$

The remaining allowance budget that an agent has at its disposal at time t is divided by the remaining days for which the allowances were issued, in order to calculate a daily budget. This budget is subtracted from the allowance quantity needed for power generation, thus resulting in the bid quantity that agent i submits to the CO2MarketOperator. In consequence, if an agent's budget for the current day is larger than its need for allowances, its bid quantity becomes negative, which corresponds to a selling bid. It is assumed that the market for CO_2 allowances is fully competitive, and the industries outside the electricity sector determine the market price. Generator agents submit price-independent bids, i.e. they are price-takers on the allowance market.

$$q_{t,i}^{CO2M} = q_{t,i}^{CO_2,emit} - \frac{q_{t,i}^{CO_2,bud}}{365-t-1} \tag{5.9}$$

The remaining allowance budget is updated at the end of each trading day for the following day. The amount of allowances used for emitted CO_2 quantities $q^{CO_2,emit}$ are subtracted from the current budget, and resulting trading volumes $q^{CO2M,res}$ (positive for bought allowances, negative for sold volumes) are added to it. Agent i's budget at time $t = 0$ is the sum of initial allowance quantities issued in year y for all power plants that agent i owns.

$$q_{t+1,i}^{CO_2,bud} = q_{t,i}^{CO_2,bud} - q_{t,i}^{CO_2,emit} + q_{t,i}^{CO2M,res} \tag{5.10}$$

$$\text{with } q_{0,i}^{CO_2,bud} = \sum_{g=i}^{G_i} q_{g,y}^{CO_2,init}$$

Agents do not act strategically on the market for CO_2 emission allowances – they do not develop bidding strategies through reinforcement learning. However, the costs incurred from allowance prices influence trading strategies on the electricity markets, as specified in the following section.

---

[7] If the deployment of minute reserve becomes necessary, this would occur one day after trading on the balancing power market. For simplicity, it is assumed that minute reserve quantities actually deployed are already known on the trading day. In real-world practice, minute reserve is hardly ever needed for frequency regulation – around 2% of procured capacity is actually deployed (Bundesnetzagentur, 2006) – so these quantities are negligible.

5.1.5 Learning Reinforcements and Market Interrelations

The three markets that form the electricity sector simulation model are interrelated through the agents' bidding strategies. A power generator has to decide whether to bid his generating capacity on the day-ahead or on the balancing power market (for those plants that fulfill the technical requirements to deliver minute reserve), and has to trade off between these two options. After market clearing on the first market, an agent can bid his remaining unsold capacity on the second market. Through varying the bid quantity on the day-ahead market, agents can influence and optimize their joint strategy on both electricity markets.

While optimizing their supply bids, agents consider opportunity costs c^{opp} that they could have achieved on the other market if they had sold their capacity there; these have been introduced in Sect. 2.3. Prices for CO_2 emission allowances are also included into the reinforcement as opportunity costs. A generator would always have the opportunity to solely sell allowances, thereby realizing a profit. Consequently, it aims at attaining a profit at least as high as it could have achieved through selling allowances.

Agents learn strategies separately for the day-ahead and for the balancing power market. In the implementation, they have individual instances of the learning algorithm for each of the two markets. Moreover, strategies for each bidding block on the balancing power market and for each hour on the day-ahead market are learned separately. Opportunity costs between the two electricity markets only occur for power plant types that are suitable for delivering minute reserve. Power output from all other plants can consequently only be bid on the day-ahead market, and opportunity costs from the balancing power market do not occur for these plants.

Reinforcements R that are fed back to the update function of the learning algorithm are calculated for both power markets after all markets are cleared on the current trading day. They are based on the profit π earned on the respective market. In order to facilitate setting initial propensities/Q-values for the learning algorithms, reinforcements are set relative to the maximum possible profit that an agent could have earned on the market π^{max}. Consequently, $0 \leq R \leq 1$ for all reinforcements and initial propensities/Q-values are set to $q_0 = 1.0$. The maximum possible profit depends on the variable cost c^{var} (see Table B.2 for parameter values) of the power plant deployed, so it is an individual value for each power plant.

Most generator agents own a portfolio of power plants over which they can maximize profits. At the same time, bids are set and learned separately for each plant. Consequently, the reinforcement for learning the bidding strategy of one plant should contain some information about the performance of the whole portfolio. In the model, the influence of the whole portfolio profit on the reinforcement is set through the *portfolio integration* parameter ψ. Throughout the simulations presented in this chapter, this parameter is set to a value of $\psi = 0.5$. This means that half of the reinforcement is defined through the payoff earned from the power plant for which the strategy is learned, and half of it is defined through the performance of the whole portfolio.

5.1 Design of the Simulation Model

In the case of Erev and Roth reinforcement learning, it is assured that reinforcements are always positive, through subtracting the minimum possible profit π^{min}. The reinforcement for the day-ahead market is defined as follows:

$$R^{DAM}_{h,i,g} = (1-\psi) \cdot \pi^{DAM}_{h,i,g} / \pi^{DAM,max}_{i,g} + \psi \cdot \frac{\sum_{G_i} \pi^{DAM}_{h,i,g} / \pi^{DAM,max}_{i,g}}{G_i} \qquad (5.11)$$

$$\text{with } \pi^{DAM}_{h,i,g} = q^{DAM,sold}_{h,i,g} \cdot P^{DAM}_h - c^{var} - c^{BPM,opp}_{h,i,g} - c^{CO2M,opp}_{h,i,g}$$

$$\pi^{DAM,max}_{i,g} = q^{net}_{i,g} \cdot P^{DAM,max} - c^{var}$$

$$c^{BPM,opp}_{h,i,g} = I(g,\text{MR}) \cdot q^{DAM}_{h,i,g} \cdot P^{BPM,cap}_{k(h)}$$

$$c^{CO2M,opp}_{h,i,g} = q^{CO_2,emit}_{h,i,g} \cdot P^{CO2M}_{t(h)}$$

The reinforcement for trading on the balancing power market is defined in a similar manner as on the day-ahead market. If no minute reserve energy is actually deployed for frequency regulation, profits are only defined by capacity prices and no-load costs (*nolc*) over the number of hours per bidding block $|h \in k|$, i.e. the cost for keeping the power plant in a stand-by state. If minute reserve is deployed, no-load costs do not occur and the profit gained from the energy price minus the variable cost has to be added. For simplicity, only the first of the two cases is formulated below. For the calculation of $\pi^{BPM,max}$, only capacity prices are considered.

$$R^{BPM}_{k,i,g} = (1-\psi) \cdot \pi^{BPM}_{k,i,g} / \pi^{BPM,max}_{i,g} + \psi \cdot \frac{\sum_{G_i} I(g,\text{MR}) \cdot \pi^{BPM}_{k,i,g} / \pi^{BPM,max}_{i,g}}{\sum_{G_i} I(g,\text{MR})} \qquad (5.12)$$

$$\text{with } \pi^{BPM}_{k,i,g} = q^{BPM,cap,sold}_{k,i,g} \cdot P^{BPM,cap}_k - c^{nolc} \cdot |h \in k| - c^{DAM,opp}_{k,i,g}$$

$$\pi^{BPM,max}_{i,g} = q^{net}_{i,g} \cdot P^{BPM,cap,max} - c^{nolc} \cdot |h \in k|$$

$$c^{DAM,opp}_{k,i,g} = q^{BPM,cap}_{k,i,g} \cdot \sum_{h \in k} P^{DAM}_h$$

5.1.6 The Graphical User Interface

In an attempt to make the electricity sector simulation model usable and understandable for decision makers in the industry, a graphical user interface has been developed. The GUI simplifies the definition of simulation scenarios and delivers plots and graphics that illustrate simulation outcomes. The tool bar, plots and other graphical features delivered by Repast are integrated into the developed GUI.

The simulation scenarios that can be defined with the help of the GUI are stored in the *Extensible Markup Language* (XML) format. Existing XML files that describe simulation scenarios can also be read into the GUI for further processing. The GUI does not set parameter values directly, but produces the XML files that are used as

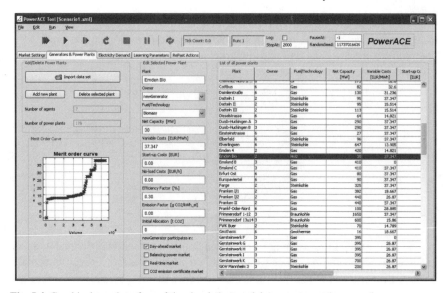

Fig. 5.4 Graphical user interface of the simulation model: generator settings panel

inputs for the simulation model. This is an important and useful functionality that ensures that simulation runs can easily be reproduced, because all relevant settings (agents and their power plants, parameters for the reinforcement learning algorithm, demand agents and also the random number seeds) are stored outside the simulation program.

Figure 5.4 shows a screenshot of the graphical user interface for the agent-based electricity simulation model. On the tab shown, the power plants to be applied can be defined through inserting or deleting single plants and setting their parameter values individually. Given datasets that define whole power plant portfolios can also be integrated through an import function.

On further tabs (see Fig. 5.5), one can define which markets are executed at which time during the simulation and what market mechanisms they apply (a). The demand side (b) can be set as an hourly system's total load or as actively bidding agents. The user can access UCTE load data or standard daily load profiles of different consumer categories (households, agriculture, business, delivered by VDEW, 1999) for defining the demand side of the simulation. The settings panel for learning algorithms (c) allows to define the representation of an agent's adaptive behavior. All learning models presented and discussed in Chap. 4 can be chosen, either globally for all agents or individually for every single participating agent. The graphical outputs that can be created with Repast are displayed on an additional tab of the GUI (d). The main menu offers common functionality such as dialogs for opening and saving files, choice of the plots that the user wishes to be shown, an undo and redo function, among others.

5.2 Validation

(a) Settings of involved marketplaces (b) Demand settings

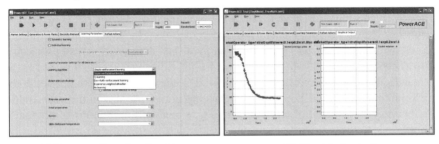

(c) Settings of learning algorithms (d) Graphical output

Fig. 5.5 Graphical user interface of the simulation model: further settings panels

5.2 Validation

"Conceptually, if a simulation model is 'valid', then it can be used to make decisions about the system similar to those that would be made if it *were* feasible and cost-effective to experiment with the system itself". (Law, 2007).

The aim of the agent-based power market simulation developed here is to apply it to policy making questions, i.e. to distinguish between market structures and market designs that make it more or less difficult for participants to drive prices away from competitive levels. If the simulation model is used to investigate such research questions, it must be assured that it is a *valid* representation of the real-world system. The measures of performance used for validating should include those that decision makers in the electricity industry will also use for evaluating system designs. In the validation procedure conducted here, it is assessed whether the order of magnitude of simulated prices on the day-ahead electricity and on the balancing power market correspond to those observed in real-world markets, and whether characteristic real-world daily and seasonal courses of prices can be reproduced by the model.

The analysis of learning algorithms provided in Chap. 4 constitutes the *micro-validation* of the simulation model. Validation on the micro level requires testing which behavioral representation is best suited for modeling agents engaged in daily repeated electricity trading. Using the learning models that came out of this analysis, the next validation step can be carried out.

During the *macro-validation* procedure, macro variables that result from the interaction of the agents participating in the modeled markets are analyzed. Here, market prices from simulations with the electricity sector model, run with data input that represents the German electricity sector, are compared to empirically observed prices.

5.2.1 Simulation Results of the Reference Scenario

The presented simulation model is run with data input that characterizes the German electricity industry during the analysis period of 2006. The power plant portfolio is represented in an aggregate way, based on published data about installed capacity. The system's total load data provided by UCTE constitutes the data input for the demand side. Simulation results are compared to empirically observed prices at the EEX spot market and at the German balancing power markets in the year 2006.[8] Furthermore, the impact of emissions trading is analyzed in order to assure that it corresponds to the real-world characteristics.

The four dominant players in the market (E.ON AG, RWE Power AG, Vattenfall Europe AG and EnBW Kraftwerke AG) are represented in more detail, and further players are introduced so that the overall installed capacity and the proportions of different generating technologies (coal-fired, gas-fired, hydro, etc.) are properly represented. Capacities and variable generating costs are defined on the basis of data published in BMWI (2007), in EWI/Prognos (2007) and in handouts issued by the generating companies. Within the power plant portfolio of one generator, all plants using the same fuel/technology are subsumed under one generating unit, and average efficiencies are assumed for these units (the average efficiency of all currently deployed power plants is 37% for lignite-fired plants, 40% for hard coal power plants, 42% for oil-fired plants and for generating units firing natural gas (Machat & Werner, 2007); no differentiation is made between gas-fired combined cycle generators and steam turbines). The resulting supply side scenario is summarized in Appendix B.2. A merit order of all power plants is depicted in Fig. 5.6.

The system's total load is regularly published by the UCTE for every third Wednesday of a month.[9] This load data for all months of 2006 constitutes the data input of the demand side for the simulations of the scenarios presented here. The hourly load for 2006 is depicted in Fig. 5.7; a table with all numbers is also provided in Appendix B.1.

All simulations are run over 7,300 iterations. As electricity trading simulations are not terminating, the outcome of one run is defined as the steady-state value of

---

[8] The year 2006 has been chosen as the analysis period because the emissions trading scheme was already fully in place in this year and data about load, prices and installed capacity for 2006 was available at the time of writing this thesis.

[9] Data can be accessed at http://www.ucte.org/services/onlinedatabase/consumption/

5.2 Validation

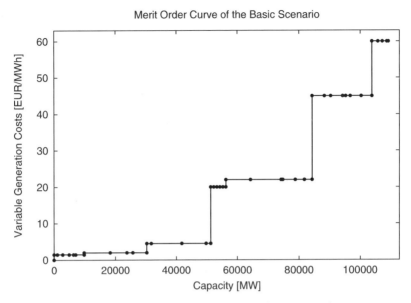

Fig. 5.6 Merit order of installed generating capacity for the reference scenario

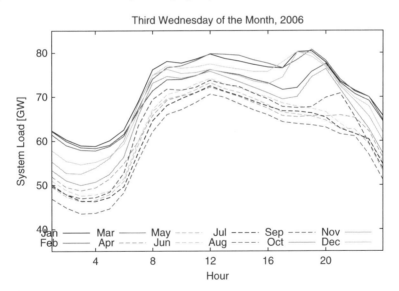

Fig. 5.7 UCTE load for 2006

one run, which is an average over the k last iterations. If the resulting prices at all iterations of run j are denoted $x_{j1}, x_{j2}, ..., x_{jm}$ with $m = 7,300$, this implies that the result of run j is equal to $\frac{1}{k} \sum_{i=m-k+1}^{m} x_{ji}$; here $k = 365$.[10]

---

[10] The length of one run and of the interval over which resulting prices are averaged to form the outcome of this run has been defined on the basis of a procedure that resembles Welch's graphical

Due to the stochastic nature of reinforcement learning, simulations have to be repeated several times with different random number seeds at each run, in order to prevent that outcomes are biased by the applied random number sequence. Here, ten repetitions are made for each simulation setting. In order to avoid unintended similarity of agents, each one has its own random distribution, which is initialized with a random number taken from the main random number distribution. Through this procedure, every agent's learning algorithm uses different random numbers, but the simulation is still fully reproducible if the applied random number seed is stored. The outcome of one simulation scenario is represented by the average over the ten repetitions for this scenario.

For the comparison of simulated and real-world prices of the reference scenario, those days for which the system's total load is known from UCTE data are simulated, and resulting prices are compared to EEX and balancing power market prices. As the real-world markets may show extraordinary prices on the specific simulated day, additional average daily courses of prices over all workdays of the same month are calculated and compared to the simulation outcomes. Figures B.1 through B.9 display simulation results for runs with Q-learning, with the original Erev and Roth reinforcement learning algorithm, and with the modified Erev and Roth algorithm (see Chap. 4 for a discussion of the learning models applied in the simulation model). Two examples are also displayed in Figs. 5.8 and 5.9 (both from simulations with Q-learning with ε-greedy action selection; learning parameter $\alpha = 0.5$, discount rate $\gamma = 0.9$ and exploration parameter $\varepsilon = 0.2$). In these figures, continuous lines plot the simulation outcome for the third Wednesday of the respective month; dashed lines plot the empirically observed prices of the same day, and dotted lines represent average prices over all workdays of the specific month. Left figures display hourly results on the day-ahead market, where empirically observed prices correspond to prices for hourly contracts fixed in the daily EEX spot auction. Right

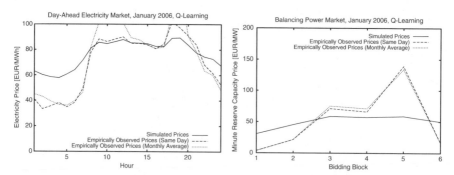

Fig. 5.8 Simulated and empirically observed prices at the day-ahead and balancing power market, January 2006, with Q-learning

procedure as described in Law (2007). All considered simulation runs have stabilized to a steady-state value at the latest after $(7,300 - 365 + 1)$ iterations, so the average over the last 365 iterations can safely be regarded as the steady-state value of one simulation run.

5.2 Validation

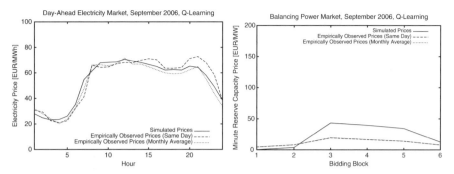

Fig. 5.9 Simulated and empirically observed prices at the day-ahead and balancing power market, September 2006, with Q-learning

figures show capacity price results from the simulated balancing power market, and the empirically observed prices are averaged over the capacity prices published by the four balancing market operators.[11]

The simulated prices observed on the day-ahead market and on the balancing power market stem from the same simulation run and are a consequence of agents bidding jointly on these two markets (and in addition on the market for CO_2 emission allowances) and optimizing their strategies in face of market interrelations.

The demand on the balancing power market, i.e. the tendered minute reserve quantity, is equal for all bidding blocks. This market is cleared first, and the day-ahead market is operated subsequently. As the available supply capacity and the demand quantity in the balancing power market is the same in every hour, differences in prices between the bidding blocks can only result from the inclusion of opportunity costs in the agents' reasoning. The simulation outcome on the balancing power market shows characteristic daily courses of prices, in which capacity prices in bidding blocks 3 and 4 – and 5 in winter months – are considerably higher than those in the nocturnal bidding blocks. Similar characteristics can be observed in the real-world balancing power markets in Germany, although the high prices in the fifth bidding block that occur in most winter months can not be reproduced by the simulation model. It is remarkable that the rather low capacity prices in some summer months can be reproduced by the simulation although the possible capacity bid prices that range up to 200 EUR/MWh would theoretically allow much higher prices to occur. This result strengthens confidence in the model validity.

Prices in summer and transitional (spring/fall) months (see example in Fig. 5.9) are quite closely reproduced by the simulation model, with the exception of July

---

[11] During the period of analysis (2006), bidding blocks were defined differently in the four balancing power markets existing in Germany. Two markets (E.ON and EnBW) solely had the two bidding blocks *high tariff* and *low tariff*; the other two TSOs had defined blocks of four to eight hours duration. From the prices on the four markets at that time, average prices for the bidding blocks defined in the simulation are calculated and – where necessary – interpolated. Monthly average prices are calculated for workdays only. Since December 2006, all four TSOs operate one joint minute reserve procurement auction with six equally sized bidding blocks: 0–4 am, 4–8 am, 8–12 am, 12 am–4 pm, 4–8 pm, and 8–12 pm.

2006. Prices in winter months (see example in Fig. 5.8), however, show larger deviations between simulation outcomes and real-world prices. In some hours, prices even exceed 100 EUR/MWh; as the agents' action domains in the simulation model only allows for bid prices up to 100 EUR/MWh, market prices above this level cannot be replicated with the model by construction. Enlarging the action domain might be a way to ensure that prices higher than 100 EUR/MWh can also be simulated. However, a variation of the range of possible bid prices also has other implications on simulation outcomes, as discussed in the sensitivity analysis (Sect. 5.2.2).

In the simulation model, prices are mainly influenced by the demand level, as the principal difference of market conditions in the hours of the considered months is the system's total load. Power plant availability is considered to be constant over the year. This is a simplification which might be altered in future model development. In reality, maintenance of power plants is scheduled discontinuously over the year; around 2% of the total installed generating capacity is off due to maintenance during winter months, and around 10% during summer months (VDN, 2004). In those simulated hours in which day-ahead electricity prices deviate considerably from real-world prices, power plant availability may be an important reason. Besides maintenance, an even more important factor in this context is the available renewable energy production. In the simulation model, renewable energy availability is also assumed to be constant, whereas in reality, water levels of hydroelectric installations and electricity generation from wind energy varies considerably throughout the year and during the day. The high prices in July 2006, which can not be replicated by the simulation model, are also explicable by reduced power plant availability. During the very hot summer in Germany in 2006, it occurred that the maximum admissible temperature for rivers was reached and the cooling water flow for thermal power plants had to be reduced as a consequence. Additional drought in many European regions reduced hydro energy availability (EGL, 2006). The combination of these factors, which were not represented in the simulation model, made power prices rise considerably above usual levels in July and, to a lower extent, August 2006.

Variability between different runs (i.e. runs with different random number seeds) is very low for simulations with Q-learning with ε-greedy action selection (see Sect. 5.2.2 for results from simulations applying other learning models). The standard deviation for the resulting prices of the ten repetitions ranges between 0.2 and 2.3 EUR/MWh for different hours on the day-ahead electricity market and between 0.05 and 3.9 EUR/MW for bidding blocks on the balancing power market. With these low variances, one single simulation run already delivers meaningful and reliable results.

It is difficult to apply statistical methods for the comparison of real-world and simulated data. As the model of a complex system such as wholesale electricity trading can only be a rough approximation of reality, the null hypothesis that distributions or mean values for real-world observations and model data outputs are the *same* is most likely false for all models. However, it is important to test whether the differences between the system and the model are significant enough to affect any conclusions derived from the model. Law (2007) proposes several methods for

5.2 Validation

comparing real-world observations with simulation output data. The basic idea behind one of these methods, the *correlated inspection approach*, has already been followed in the preceding paragraphs: the model has been driven with historical data input (i.e. historical system's total load values), and outputs have been graphically compared for all hours and bidding blocks (see plots in Appendix B.4). It has been shown that simulation data corresponds well to real market data in numerous cases; in some cases, however, deviations between these two are not negligible.

In order to get a measure of whether the overall picture of this model outcome can still be regarded as a valid representation of the real system, it is next attempted to apply the *confidence interval approach* described by Law (2007) to the data: if X_j is a random variable defined by system data, and Y_j is the same variable defined by model data, then the confidence interval approach compares the model with the system by constructing a confidence interval for $\zeta = \mu_X - \mu_Y$, where $\mu_X = E(X_j)$ and $\mu_Y = E(Y_j)$ correspond to the mean of the respective data set. A paired-t approach can be used to construct the 100(1-α) confidence interval for ζ. If $l(\alpha)$ and $u(\alpha)$ are the corresponding lower and upper confidence interval endpoints, respectively, and if $0 \notin [l(\alpha), u(\alpha)]$ then the observed differences between μ_X and μ_Y are statistically significant at the level α. The average values over n observations are unbiased estimators of the mean values μ_X and μ_Y:

$$\overline{X}(n) = \frac{\sum_{j=1}^{n} X_j}{n} \text{ and } \overline{Y}(n) = \frac{\sum_{j=1}^{n} Y_j}{n}$$

For simulations using Q-learning, if X_j's correspond to the observed hourly prices at the EEX market, and Y_j's are (average) simulated hourly prices of the same days, $\overline{X}(n) - \overline{Y}(n)$ is 0.13 EUR/MWh over the 288 simulated hours on the day-ahead market.[12] The 90% confidence interval for $\zeta = \mu_X - \mu_Y$ is 2.30, so given that the interval [−2.17, 2.43] contains 0, the observed differences between μ_X and μ_Y are not statistically significant. For the balancing power market, $\overline{X}(n) - \overline{Y}(n)$ results in −16.29 EUR/MW; the 90% confidence interval is 17.41. Here, too, the interval [−33.4, 1.12] contains 0, so differences between simulation outcomes and real-world data can be regarded as not statistically significant.

Although results from this calculation confirm the model quality, it must be noted that the applicability of the confidence interval approach to the data analyzed here is restricted: this method requires that the sets of X_j and Y_j are both independent. In the real-world, however, hourly prices can usually not be regarded as independent from each other. Besides, the methodology mainly gives evidence about how good mean (yearly) prices are replicated by the model. Here, however, it is also interesting whether characteristic daily price curves are replicated, so the evidence given by the confidence interval approach is only limited for the purpose of validating the electricity simulation model.

---

[12] 24 h for each of the twelve months results in 288 observations on the day-ahead market. On the balancing power markets, six daily bidding blocks for 12 months result in 72 observations.

From the discussion provided here, it becomes noticeable that it is difficult to judge the validity of the model results. It is argued here that model validity is sufficient for analyzing the research questions presented in Chap. 6, because characteristic courses of prices (with high prices around noon and in the evening, and lower prices during the night), and the overall order of magnitude of real-world prices have been successfully replicated through simulation runs.

5.2.2 Sensitivity Analysis

Among the model parameters that have to be set to certain values, some may have a considerable influence on simulation results. For an enhanced understanding of the model characteristics, it is important to find out how these parameters effect simulated market prices. This is accomplished in the sensitivity analysis reported here. Sensitivity is tested for the learning algorithm applied for representing agent behavior, for the absolute range of possible actions forming the action domain, and for the portfolio integration parameter ψ. Finally, the impact of emissions trading is analyzed in order to test how the introduction of the market for CO_2 emission allowances influences electricity market prices.

In analogy to the simulations with the reference scenario reported in the previous section, ten simulation repetitions with different random number seeds have been run for every setting. In order to reduce avoidable variance between runs, the method of *common random numbers* (Law, 2007) is used. A list of ten random number seeds is given as an input to the simulation model; at each repetition, the program takes the next seed from this list and initializes the main random number distribution of the model with this seed. Through this procedure, it is also easy to reproduce simulation runs on the basis of the list of random number seeds given as data input to the model.

If the series $x_{11}, x_{12}, ..., x_{1m}$ is one realization of the random variables $X_1, X_2, ..., X_m$ resulting from a simulation run of length m (here $m = 7,300$) using a specific set of random numbers, and $x_{21}, x_{22}, ..., x_{2m}$ up to $x_{n1}, x_{n2}, ..., x_{nm}$ (here, $n = 10$) are further realizations of the random variables resulting from runs with different sets of random numbers, then $x_{1i}, x_{2i}, ..., x_{ni}$ are *independent and identically-distributed* (IID) observations of the random variable X_i, for $i = 1, 2, ..., m$ (Law, 2007). Due to the independence across runs, the comparison of the outcomes of different settings for the simulation model can be based on simple statistical methods, which will be applied to the sensitivity analysis and for the scenarios presented in Chap. 6.

Throughout the sensitivity analysis, the *paired-t confidence interval* method (Law, 2007) is applied for comparing results from different settings. With the paired-t confidence interval, it can be tested whether results from two settings differ from each other in a statistically significant manner. If for $i = 1, 2$ the sample of n_i IID observations $X_{i1}, X_{i2}, ..., X_{in_i}$ constitutes the output of simulation runs for two different settings, then $Z_j = X_{1j} - X_{2j}$ for $j = 1, 2, ..., n$ expresses the differences between the outcomes of the two settings. The Z_j's are also IID random variables,

5.2 Validation

and the confidence interval is constructed for $E(Z_j) = \zeta$. The $100(1-\alpha)$ percent confidence interval for ζ is

$$\bar{Z}(n) \pm t_{n-1,1-\alpha/2}\sqrt{\widehat{\text{Var}}[\bar{Z}(n)]} \tag{5.13}$$

with the sample mean

$$\bar{Z}(n) = \frac{\sum_{j=1}^{n} Z_j}{n} \tag{5.14}$$

and the variance of the mean

$$\widehat{\text{Var}}[\bar{Z}(n)] = \frac{\sum_{j=1}^{n}[Z_j - \bar{Z}(n)]^2}{n(n-1)} \tag{5.15}$$

Recall that here, $n_1 = n_2 = 10$, i.e. ten repetitions have been run for every setting.[13] Following the *common random numbers* method, the pairs of X_{1j}'s and X_{2j}'s are the outcomes of those simulation runs of the two settings that use the same random number seed. If the calculated confidence interval does not contain zero, the null hypothesis H$_0$: $\mu_{X_1} = \mu_{X_2}$ is rejected and differences between the two simulation settings can be regarded as statistically significant; in addition, the confidence intervals gives an indication of the magnitude by which the two means differ from each other. If the interval contains zero, the null hypothesis can not be rejected, i.e. the two means can not safely be regarded as different from one another. The paired-t confidence interval method is applied to all observations (hours/bidding blocks) that are simulated for each setting, i.e. to the 288 points of measurement on the day-ahead market and to 72 points of measurement on the balancing power market.

Varying the Learning Algorithm

Following the postulate formulated in Chap. 4, simulation results should always be confirmed by comparative runs applying different learning models. If several learning models are deemed appropriate for representing agent behavior (as it was found in this model for the cases of Q-learning and the original and modified Erev and Roth reinforcement learning algorithms), simulation outcomes should be considered as valid only if they are confirmed with all of these learning models.

Simulation results for the reference scenario have been reported for runs in which Q-learning is applied as the agents' behavioral representation. In Appendix B.4, hourly simulated prices compared to real-world prices are also depicted for

---

[13] It should be noted that in contrast to the *confidence interval approach* for comparing simulation outcomes with real-world data, which has been applied earlier in this section, here, the different outcomes of n repetitions of the same simulation scenario are used as the basis for comparison. The first approach was applied to the average values over the n repetitions, and 288 of these hourly (average) values were compared to the 288 hourly prices of the real-world system.

simulations with the Erev and Roth learning algorithm: for simulations using original Erev and Roth reinforcement learning, see Figs. B.4, B.5, and B.6, and for simulations applying the modified Erev and Roth RL, see Figs. B.7, B.8, and B.9.

In order to facilitate the graphical inspection of simulation results, the following plots contain resulting prices for all simulated observations (12*24 points on the DAM and 12*6 points on the BPM). As prices on both electricity markets are strongly influenced by the system's total load (= demand), simulated prices are sorted by load quantities in the corresponding hours (DAM), or by average load over the four hours constituting a bidding block (BPM). System load is plotted at the second ordinate of the diagrams. The resulting curves are depicted in Figs. 5.10, 5.11, and 5.12 for the three learning variants considered.

The simulations with Erev and Roth reinforcement learning apply the proportional action selection rule with a recency value of $\phi = 0.1$ and an experimentation value of $\varepsilon = 0.2$, those with Q-learning have the same settings as the simulations presented in Sect. 5.2.1. Further runs with both Q-learning and Erev and Roth learning applying the Softmax action selection rule have been conducted. Prices in these scenarios deviate stronger from real-world prices than those with proportional or ε-greedy action selection, so these are not further reported here. Also, a variation of

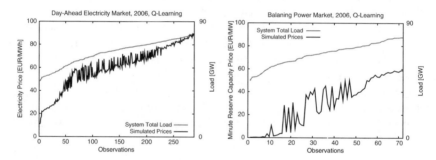

Fig. 5.10 Prices for all observations at the day-ahead and balancing power market, sorted by corresponding system load (Q-learning simulations)

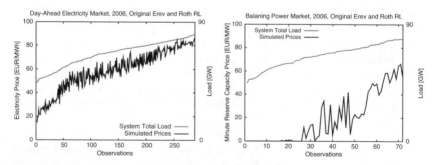

Fig. 5.11 Prices for all observations at the day-ahead and balancing power market, sorted by corresponding system load (original Erev and Roth RL simulations)

5.2 Validation

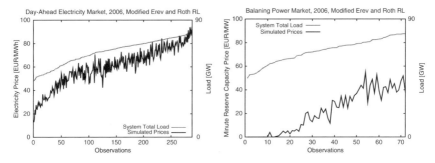

Fig. 5.12 Prices for all observations at the day-ahead and balancing power market, sorted by corresponding system load (modified Erev and Roth RL simulations)

the discount rate and exploitation rate for Q-learning has been tested, but does not result in prices that are closer to observed prices in the real-world.

Simulation results vary considerably between the ten repetitions for simulations applying Erev and Roth reinforcement learning. The standard deviations of the single observations are between 3.3 and 13.0 EUR/MWh on the day-ahead and 0.0 and 29.0 EUR/MW on the balancing power market when applying the original algorithm, and 3.4–17.9 EUR/MWh (DAM) and 0.0–24.5 EUR/MW (BPM) for simulations with the modified Erev and Roth algorithm.

In order to get a measure of how close simulated prices with Erev and Roth RL are to real-world prices, the confidence interval approach (see Sect. 5.2.1) has been applied for results from these simulations, too. Good results are found for simulated prices on the day-ahead market; on the balancing power market, however, differences between simulated and real-world prices are statistically significant. In the following, only the resulting intervals are given: [−0.66,3.94] (DAM, original Erev and Roth); [10.44,44.80] (BPM, original Erev and Roth); [−2.01,2.65] (DAM, modified Erev and Roth); [7.58,42.38] (BPM, modified Erev and Roth).

The paired-t 90% confidence intervals have been calculated for all observations of the day-ahead and balancing power market. The confidence intervals are constructed around the difference in two observation means, as described above in this section. Here, the difference between results from simulations with Erev and Roth learning and with Q-learning are analyzed in order to judge how strongly they deviate from each other. Results show that for the majority of simulated hours, mean results from Q-learning simulations do not differ significantly from results of simulations with the original or modified Erev and Roth reinforcement learning algorithm. In about 20% of all simulated hours on the day-ahead market, however, differences in simulated electricity prices are statistically significant. Lower and upper endpoints of the confidence intervals on both power markets are depicted in Figs. 5.13 (original Erev and Roth RL) and 5.14 (modified Erev and Roth RL). It can also be seen that on the balancing power market, simulation results deviate stronger for different applied learning models. Especially in hours of low system's total load, mean simulated prices with Erev and Roth learning are significantly lower than in simulations with Q-learning.

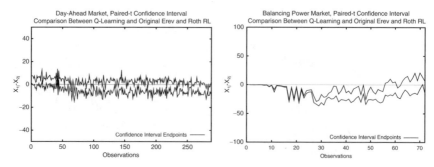

Fig. 5.13 90% confidence intervals for $E(Z_j)$, comparison of Q-learning and original Erev and Roth RL

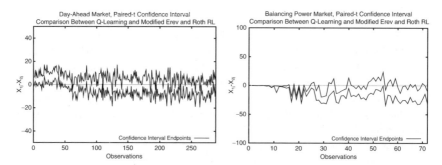

Fig. 5.14 90% confidence intervals for $E(Z_j)$, comparison of Q-learning and modified Erev and Roth RL

The comparison of Q-learning and Erev and Roth learning leads to the conclusion that the quality of results, i.e. similarity to real-world prices and variance between runs, is better for simulations applying Q-learning. But as simulations with Erev and Roth RL also produce characteristic daily price curves and the order of magnitude of real-world prices, simulations with this algorithm can be used to check whether the characteristics found from simulations with Q-learning are confirmed by Erev and Roth simulations or not. If they confirm results, these have a better support; if not, the results can not be considered as sure enough for deriving policy advice from it, and additional analyses should be carried out.

Varying the Action Domain

It has been shown already for the simplified learning model test scenarios (Sect. 4.3.3) that the definition of the action domain has a noticeable effect on the level of simulated prices. This is surely a drawback of agent-based simulation models applying reinforcement learning. The absolute range of actions that an agent can choose from (here especially the range of possible bid prices) has to be carefully calibrated, so that the order of magnitude of simulated macro variables corresponds to real-world values.

5.2 Validation

The influence of the range of day-ahead bid prices is tested here in order to discover the characteristics of its impact on resulting prices for day-ahead electricity contracts and for minute reserve on the balancing power market. The reference value of maximum possible day-ahead bid prices $p^{DAM,max}$ is 100 EUR/MWh. Four alternative scenarios with maximum prices of 90, 110, 120 and 150 EUR/MWh have been simulated and compared in the framework of this sensitivity analysis.[14] In Figs. 5.15 and 5.16, simulated hourly electricity prices on the day-ahead and prices per bidding block on the balancing power market are plotted in order of increasing system load of the corresponding observation, for the four listed alternative scenarios and for the reference scenario.

An enhanced understanding of the influence of the action domain specification on prices can be gained from Fig. 5.17, where relative prices in reference to the scenario of $p^{DAM,max} = 100$ EUR/MWh are plotted (for every observation of one scenario, the resulting price is divided by the corresponding price of the reference scenario). From the analysis, three characteristics can be observed:

- The admissible bid price range clearly influences resulting prices; on the day-ahead market, the decrease or increase is nearly proportional, i.e. the quotient of prices in one scenario over the reference scenario is more or less constant. Only for the very wide action domain with bid prices of 0–150 EUR/MWh, stronger relative deviations from the reference case are observable for very low and for high demand hours.

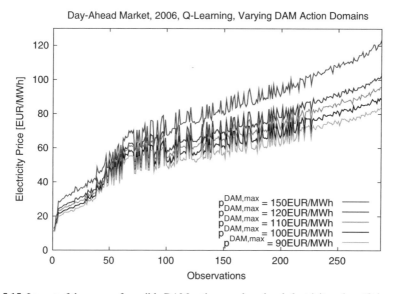

Fig. 5.15 Impact of the range of possible DAM actions on day-ahead electricity prices (Q-learning simulations)

---

[14] It should be noted that the number of actions is kept constant over all scenarios, i.e. 21 discrete price steps are defined between the minimum and maximum possible bid price.

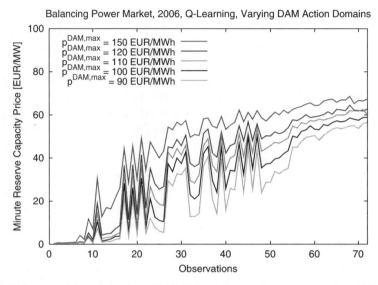

Fig. 5.16 Impact of the range of possible DAM actions on minute reserve prices (Q-learning simulations)

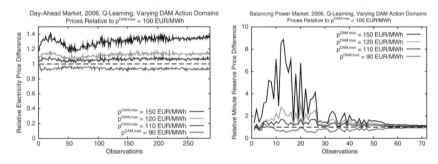

Fig. 5.17 Relative differences of simulated prices with varying DAM action domains

- The relative decrease or increase in day-ahead prices is less than the decrease or increase in maximum admissible bid prices on the same market. If the latter is augmented to 110, 120 or 150 EUR/MWh, average yearly prices rise by 6.2%, 12.5%, and 31.3%, respectively. If it is lowered to 90 EUR/MWh, yearly average prices decrease to a level of 93.8% of those in the reference scenario.
- On the balancing power market, the impact of varying action domains is less obvious. There seems to be a certain level up to which higher opportunity costs resulting from higher day-ahead electricity market prices are carried over to balancing power market results. At the high demand levels, market prices on the balancing power market hardly differ for the alternative action domain scenarios.

The paired-t confidence interval test at the 90% level confirms that simulation results are statistically significantly different from each other when the action domain is varied. Results from simulations with $p^{DAM,max} = 90$ are pairwise compared with $p^{DAM,max} = 100$, those with $p^{DAM,max} = 100$ are pairwise compared with $p^{DAM,max} = 110$, and so forth. A graphical presentation of all confidence intervals for Q-learning simulations are summarized in Fig. B.10 in Appendix B.5. On the day-ahead market, prices resulting from action domain settings which allow higher bid prices are significantly higher than prices of settings in which allowed bid prices are lower, with very few exceptions. On the balancing power market, prices for very low system's total load levels are not always statistically significantly different for two neighboring settings, but results are significantly different for the mid and high demand bidding blocks (more than 80% of all blocks).

The findings of this analysis confirm the assumption gained from results of the simplified scenario (Chap. 4), i.e. that simulation results are sensitive to the definition of the agent's action domain, which constitutes a deficiency of the agent-based methodology applying reinforcement learning.

Varying the Portfolio Integration Parameter

As specified in Sect. 5.1.5, each agent who owns a portfolio of power plants takes into account the profit earned from the specific plant for which it is learning the profit-maximizing bidding strategy, as well as the profit earned from the whole portfolio of generation assets. The portfolio integration parameter ψ was introduced to quantify the composition of the reinforcement that is fed back to a bidding strategy for a specific power plant, see (5.11) and (5.12).

If $\psi = 0$, only the profit gained by the individual power plant is considered for the reinforcement of strategies, and profits from the other generation units of the agent's portfolio are neglected. This would be equal to a situation in which each agent only owns one power plant, and corresponds to a case of stronger supply side competition, and – as a consequence – lower market prices. If part of the reinforcement is defined by the overall trading success of the whole power plant portfolio, agents have more possibilities to optimize bidding strategies over their generating portfolios. Larger agents may have more potential to exert market power through successfully bidding at higher prices, and overall market prices can be expected to be higher, too.

As players in the electricity market are reluctant to reveal their trading strategies, and given the fact that the portfolio integration parameter defined here is only a strongly simplified approximation of multi-plant trading strategies, it is not possible to get any empirically grounded value for this parameter. In the reference scenario, ψ is set equal to 0.5. This sensitivity analysis can only show how strongly variations of this value influence simulation results.

In Fig. 5.18, prices resulting from simulations with $\psi = 0.3$ and $\psi = 0.7$ are compared to the reference case of $\psi = 0.5$. The design of the plot is similar to that chosen in Fig. 5.17, in that it plots relative prices as compared to the reference case, sorted by system load of the corresponding observation; it also uses the same scale.

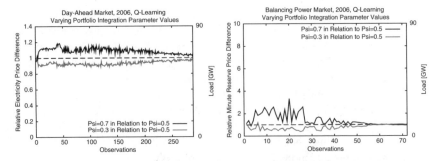

Fig. 5.18 Relative differences of simulated prices with varying portfolio integration parameter

It can be seen that a variation of the portfolio integration parameter actually has an effect on simulated prices. This effect becomes smaller in hours of high demand. Interestingly, the demand levels at which the effect of varying ψ values becomes lower (roughly, starting from observations 230) is the same as those at which the impact of emissions trading on electricity prices becomes nearly constant – this aspect will be treated in the next subsection. One possible interpretation of this pattern is that in these hours, agents have less possibilities to optimize overall portfolio profits through bidding less carbon intensive plants into the market. This can be explained by the necessity to run more carbon intensive power plants in order to satisfy demand. As soon as the possibilities of avoiding costs through bidding more capacity of gas-fired power plants and less of coal or lignite-fired plants are restrained, the parameter integration effect has less impact on trading strategies and, as a consequence, on resulting market prices. In hours of very low demand (the very first observations), competition seems to be so strong that agents cannot successfully raise prices through varying their bidding strategies, so the portfolio integration factor has no influence in these hours.

The paired-t confidence interval approach at a 90% confidence level confirms that results for the two settings with varying portfolio integration parameter values are different from the reference scenario in a statistically significant way (see Fig. B.11 in the appendix for an overview of all confidence intervals). This is true for almost all observations on the day-ahead market, except for three in which load is very low. On the balancing power market, prices are significantly different for more than 55 out of 72 observations for both tested values of ψ. For the remaining bidding blocks, no significant differences in resulting minute reserve prices can be observed; this is the case in times of high or very low system's total load (cp. Figs. 5.18 and B.11).

Impact of CO_2 Emissions Trading on Electricity Prices

The data presented in the preceding section corresponds to simulations in which emission allowance trading was integrated – just like in the real-world market of the corresponding time frame. In further simulation runs, it is tested how emissions trading affects prices on the electricity markets. For this purpose, scenarios without CO_2 emissions trading are run and compared to the reference scenario results. The outcome of this comparison is depicted in Figs. 5.19 and 5.20 for the day-ahead electricity market and for the balancing power market, respectively. It can be shown that a large fraction of opportunity costs resulting from the possibility of selling CO_2 emission allowances is successfully passed over to electricity market bids, which ultimately raises prices at the day-ahead market and also at the balancing power market. Because of different emission and competition situations in the single hours, the absolute increase in electricity prices is not constant across the simulated hours and bidding blocks.

In hours of low demand, the introduction of emissions trading has hardly any effect on day-ahead electricity prices, because only few power plants that incur high CO_2 emissions are deployed, and supply side competition is strong. In contrast, the difference in prices is considerable in high demand hours, in which many CO_2 intensive power plants are running and competition is weak, so agents can successfully pass over additional opportunity costs to their bid prices. Over a large range of intermediate demand situations, deviations between the scenarios with and without emissions trading fluctuate to some extent. The intuition behind this result is that these hours with similar demand situations belong to different months, and CO_2 prices differ across months. Hours with very high demand all belong to the winter months in which demand is high and consequently many fossil fuel power plants are operated, resulting in (evenly) higher CO_2 allowance prices. This is also illustrated by the green curves that plots prices for CO_2 allowances in Figs. 5.19 and 5.20.

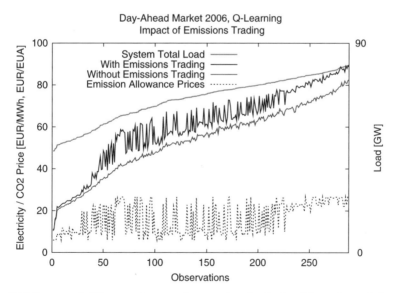

Fig. 5.19 Impact of CO_2 emissions trading on day-ahead electricity prices (Q-learning simulations)

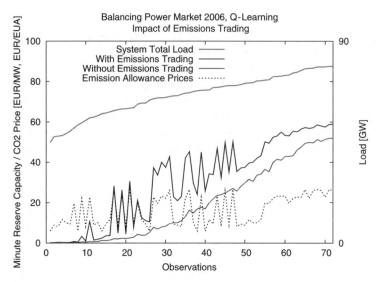

Fig. 5.20 Impact of CO_2 emissions trading on minute reserve capacity prices (Q-learning simulations)

As a consequence, it can be concluded that emissions trading considerably influences electricity prices and that it is the main cause for differences in prices resulting for hours with similar demand situations; this is true on both the day-ahead and the balancing power market. Yearly average prices are 13.3% higher for scenarios with emissions trading on the day-ahead market, and 56.8% higher on the balancing power market (for simulations with Q-learning).

The paired-t 90% confidence interval analysis confirms that resulting prices are significantly higher when emissions trading is introduced, as compared to a the test case without emissions trading (for simulations with Q-learning). This result is obtained at all observations on the day-ahead market and at all except for four low demand observations at the balancing power market. The extent to which prices deviate from one another is different across demand levels, which is also illustrated by the confidence intervals constructed around the differences. Figure B.12 in Appendix B.5 graphically depicts the confidence intervals of the comparison between the scenarios with and without emissions trading.

5.3 Summary

In this chapter, the architecture of the agent-based electricity sector simulation model was outlined in detail. Simulation results from this model, carried out for a realistic scenario of the German electricity sector, reveal that real-world prices can be reproduced well for spring, summer and fall months. In winter months, simulated

5.3 Summary

prices deviate to some extent from empirically observed prices. In these months of high system load, agents in real-world markets may have more leeway for strategic bidding than has been assumed in the model presented here. Other reasons that may explain deviations between real-world prices and simulation results have also been discussed in this chapter (mainly power plant availability).

Despite some discrepancy between system and model prices, the simulation model can be considered a valid representation of the real-world system, regarding the relevant aspects needed for the research questions analyzed in this work. Characteristic daily courses of prices and the order of magnitude of prices in simulation outcomes resemble those in real-world markets. The sensitivity analysis presented in this chapter highlights the main factors that influence model outcomes. Insights from this analysis show that both the action domain and the portfolio integration factor ψ have an impact on results and must be set carefully. The influence of CO_2 emissions trading has also been quantified through a comparison with runs that exclude allowance trading. The conclusion from this analysis is that a large fraction of costs arising from the need to provide allowances for every ton of CO_2 emitted is successively passed over to electricity market bids.

Part III
Application, Evaluation and Discussion

Chapter 6
Simulation Scenarios and Results

After having calibrated and validated the agent-based electricity sector simulation model, it can now be used for analyzing the impact of specific changes in the market structure or market rules. Results from correspondingly defined simulation scenarios may deliver helpful insights for policy makers, and advise regulatory intervention in the electricity industry. Three cases are studied in this chapter, and their implications for policy making are presented and discussed.

The influence of different market structures and market mechanisms on power trading outcomes, especially on resulting market prices, is assessed through specific simulation runs. In these simulation scenarios, only one aspect of the market structure or the trading arrangement is altered at time, while the others remain unchanged as compared to the reference scenario presented in Chap. 5, including demand patterns, initial conditions, and also the random number sequences. Consequently, differences in results can be ascribed solely to the induced changes between the simulation scenarios.

Three problems of interest in the German electricity sector are reported in the following: in Sect. 6.1, the impact of varying tendered minute reserve quantities are studied; in Sect. 6.2, resulting prices under two different settlement rules on the balancing power market are compared, and in Sect. 6.3, an altered market structure on the supply side is assessed. The policy implications that arise from the impact of the tested scenarios on day-ahead electricity and minute reserve prices are formulated in Sect. 6.4.

6.1 Impact of Tendered Balancing Capacity on Electricity Prices

6.1.1 Motivation

In accordance with many other studies about European electricity markets, an expertise by order of the German *Federation of Industrial Energy Consumers* (von Hirschhausen et al., 2007) comes to the conclusion that electricity markets in continental Europe are not structured in a sufficiently competitive way. As regards the interplay between the day-ahead electricity market and the balancing power market, they furthermore conclude: "Through subtracting capacity from wholesale trading, the balancing power market particularly offers potential for market power exertion". (von Hirschhausen et al., 2007). The agent-based electricity sector model is suitable for testing whether this assumption materializes, and, if applicable, to quantify the effect.

In real-world practice during the year 2006, the procured positive balancing power capacity was nearly 4,000 MW for primary and secondary reserves, and between 3,150 and 3,420 MW for minute reserve. In the baseline case simulated in Chap. 5, procured minute reserve capacity is set to 3,500 MW, while primary and secondary reserve procurement is not considered. It would now be interesting to observe how prices for minute reserve, and also prices on the day-ahead market change when the quantity of procured minute reserve is altered. In order to quantify the changes in prices, the amount of tendered minute reserve in the simulation model is varied from 1,000 MW to 13,000 MW for all bidding blocks, and resulting prices are compared.

Standard economic theory suggests that, ceteris paribus, the price for a good rises when demand increases. In classical electricity market models that assume no strategic bidding, the rise in price for times of high system load is directly explained by the form of the merit order (see Fig. 5.3 for an illustration): if load is high, more expensive generating units must be deployed in order to satisfy demand. A rise in prices which is solely due to a shift in the merit order is, thus, no case of market power exertion. Only if bidders can successfully raise market prices to an extent which cannot be explained by the underlying cost structure alone, it can be concluded that there exists an increased "potential for market power exertion".

In the present simulation model, the balancing power market is cleared first. Among the power plants which are technically capable of delivering minute reserve, 7,000 MW of hydroelectric capacity is available at a no-load costs of 0 EUR/MW per hour and variable (operating) costs of 1.5 EUR/MWh. Additional demand can be satisfied with the available capacity of the next expensive gas-fired power plants (with no-load costs of 100 EUR/MWh and variable costs of 45 EUR/MWh, without emissions costs) in all simulated scenarios of tendered balancing capacity. Expensive oil-fired power plants do not need to be deployed for minute reserve purposes, unless they underbid lower cost hydroelectric or gas-fired generating capacity.

It is less obvious to determine the merit order shift on the day-ahead market, because this depends on the result on the balancing power market, as the latter is cleared earlier. Under the assumption that the least-cost plants are deployed for minute reserve, the variable generating cost of the last unit necessary to satisfy demand on the day-ahead market is 4.5 EUR/MWh for a few hours of very low demand, and 20–22 EUR/MWh for the largest part of all hours; for hours of very high demand and high tendered minute reserve quantities, system marginal costs rise to 45 EUR/MWh. All these considerations exclude costs for CO_2 emission allowances.

6.1.2 Results

Simulation runs have been conducted with amounts of tendered balancing power (minute reserve) varying from 1,000 MW to 13,000 MW. As for the high number of simulations needed, the number of repetitions per scenario has been reduced to five only for this analysis.

On the basis of the assumption stated in the cited expertise, the following two hypotheses are formulated:

Hypothesis 1. *Market prices for minute reserve increase as the amount of tendered balancing capacity increases.*

Hypothesis 2. *Market prices for day-ahead electricity delivery contracts increase as the amount of tendered balancing capacity increases.*

The left side of Fig. 6.1 shows simulated prices on the day-ahead market as a function of tendered minute reserve quantities for simulations applying Q-learning. Over the simulated range, a nearly linear relation between balancing power quantities and resulting prices can be observed, which seems to confirm the effect assumed in the expertise. The slope of the curve is slightly steeper for cases with higher demand, such as in winter months, than in times of low demand; Fig. 6.1 displays the two months with the highest and lowest slope, for each market.

The effect of varying tendered minute reserve capacity is much stronger on the balancing power market. As depicted in the right side of Fig. 6.1, higher tendered capacities lead to considerably higher minute reserve prices (simulations with Q-learning). The functional relation is weakly concave. The yearly average price increase from less than 15 EUR/MWh up to more than 90 EUR/MWh corresponds to a sextupling of prices between 1,000 and 13,000 MW of tendered balancing capacity.

The fact that prices do not only decrease when tendered minute reserve quantities rise above the level of 7,000 MW suggests that the dynamics of competition on the balancing power market are more influenced by other factors, such as the supply/demand ratio or opportunity costs from the day-ahead market, than by the underlying

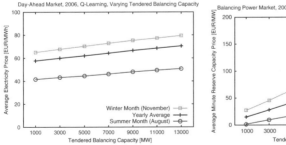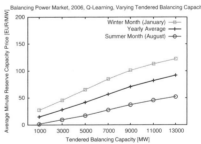

Fig. 6.1 Monthly/yearly average prices on the day-ahead (*left*) and balancing power market (*right*) for different tendered minute reserve quantities

cost structure. This can be seen as an indicator for an increasing market power potential when tendered minute reserve quantities are elevated. Higher demand on the balancing power market reduces supply-side competition, so agents are less forced to bid competitively, leading to higher market prices. Besides, if more capacity is committed on the balancing power market (which is cleared first), less capacity is available on the day-ahead market, leading to the same effects on this market, too. As opportunity costs are taken into account, high prices on one market tend to result in higher prices on the other market. This effect especially occurs on the balancing power market, as it is stronger influenced by the day-ahead market than vice versa.

Figure 6.2 shows the prices resulting on the day-ahead market for all observations (i.e. for all simulated hours), and for varying tendered minute reserve quantities. Figure 6.3 displays the resulting prices on the balancing power market. Observations are sorted by demand quantities in the corresponding hours/bidding blocks. What can be observed here is the influence of the demand level on resulting prices. Price increases resulting from elevated tendered balancing power quantities are stronger in high demand hours and have a negligible effect in hours of very low demand.

It can also be seen that in low demand hours, prices are very low and close to system marginal costs. With increasing demand, prices rise quickly and reach levels over 40 EUR/MWh in more than 80% of all observations on the day-ahead electricity market. For very high demand and high tendered balancing capacity, they even climb up close to the maximum possible price of 100 EUR/MWh. Thus, it can be concluded that agents in this simulation scenario have a substantial potential to exert market power, and that this potential is considerably augmented with increasing amounts of tendered minute reserve, especially in high demand hours.

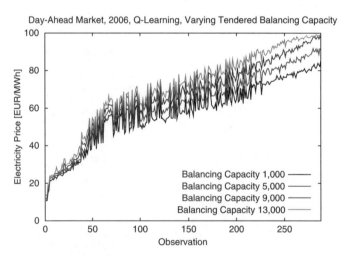

Fig. 6.2 Impact of tendered minute reserve quantities on day-ahead market prices (Q-learning simulations)

6.1 Impact of Tendered Balancing Capacity on Electricity Prices

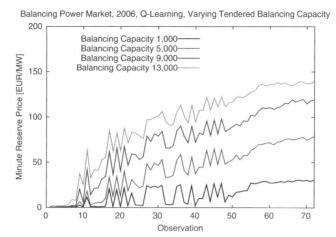

Fig. 6.3 Impact of tendered minute reserve quantities on balancing power market prices (Q-learning simulations)

A paired-t confidence interval analysis at the 90% level with simulations applying Q-learning reveals that for two neighboring scenarios, prices are statistically significantly higher in the setting with higher tendered minute reserve quantities, except for hours of very low or very high system's total load. Results from the presented simulations with Q-learning as the agents' behavioral representation confirm the assumed effect formulated in Hypotheses 1 and 2. Simulations applying Erev and Roth reinforcement learning confirm the general findings reported from Q-learning simulations. Average yearly and monthly prices on both markets also rise monotonously as a function of increasing tendered balancing capacities. The difference in average day-ahead prices between the scenario of lowest and highest tendered balancing capacity is near +20% (Q-learning: +22%, original and modified Erev and Roth RL: +18%); on the balancing power market, the resulting difference is higher in the simulations applying Erev and Roth reinforcement learning (Q-learning: +523%, original Erev and Roth RL: +940%, modified Erev and Roth RL: +750%). However, due to the higher variability between runs for simulations with the Erev and Roth algorithms, differences between two neighboring scenarios are only statistically significant for some observations. Erev and Roth simulations, thus, weakly confirm the results from Q-learning simulations. The confidence intervals around the differences of the two extreme cases with tendered balancing capacities of 1,000 and 13,000 MW, which were constructed in the same manner as those in the sensitivity analysis (see description in Sect. 5.2.2), are depicted in Appendix C.1 for all three applied learning models.

6.2 Settlement Rules on the Balancing Power Market

6.2.1 Motivation

In procurement auctions with two-part bids, the market clearing procedure distinguishes two aspects: the *scoring rule* and the *settlement rule* (Chao & Wilson, 2002). The scoring rule defines how to compare bids, and the settlement rule determines payments. Basically, two options for payments of successful bids are discussed in the literature and employed in practice: *uniform price* and discriminatory *pay-as-bid* settlement. In the former case, one market clearing price is determined – usually the price of the highest successful bid – which is paid to all successful bidders, whereas in the latter case, which is also applied in the reference scenario of the model presented here, every successful bid is paid its bid price.

In European balancing power markets, both settlement rules are employed. In Germany, Austria and the United Kingdom for example, pay-as-bid pricing is applied, whereas Sweden, Norway, Spain and the Netherlands opted for uniform pricing (Swider, 2006). The question which settlement rule is less vulnerable to strategic bidding, and which leads to lower average prices, is discussed by practitioners, policy makers and academics likewise, and with inconsistent conclusions.

In the political discussion, a naïve and intuitive argument in favor of pay-as-bid pricing is often brought forward: since all intra-marginal bids will receive more than their bid prices under uniform pricing, a shift to pay-as-bid pricing would simply eliminate those mark-ups. The average price that will be paid to the generators under pay-as-bid pricing will incorporate no mark-up above marginal costs at all.

Kahn, Cramton, Porter, and Tabors (2001) argue that uniform pricing is more appropriate for preventing inefficiencies and strengthening competition. They question the underlying assumption under the naïve argument in favor of pay-as-bid pricing, i.e. that generators will bid identically under both pricing rules. With pay-as-bid pricing, they argue instead, the bidders' incentive is to bid as close to the (estimated) clearing price as possible. The pay-as-bid auction rewards those that can best guess the clearing price. Typically, this favors larger companies that can spend more on forecasting, and are more likely to set the clearing price as a result of their size. They also argue that low-cost generators might overestimate the clearing price and bid too high, which would result in an inefficient scheduling, because more expensive units will be deployed in their place. In the absence of market power, they conclude, dispatch is likely to be less efficient under pay-as-bid than under uniform pricing.

Ausubel and Cramton (2002) advert to the argument that in multi-unit auctions like electricity auctions, uniform pricing gives generators more incentive to exercise market power, because a generator recognizes the possibility that his bid may set the clearing price he receives for the entire quantity he wins. Consequently, he is inclined to increase his bid price on some bids, which tends to result in higher market clearing prices. Under pay-as-bid pricing, in contrast, the generator's marginal bid does not impact what he is paid on the other (intra-marginal) quantities, so he would

6.2 Settlement Rules on the Balancing Power Market

be less inclined to bid at overly high prices. The authors also state that the extent of inefficiency of uniform price auctions depends on the presence of large bidders, who have an ability to exercise market power. As electricity markets are usually characterized by high concentration levels, good auction design becomes even more important.

Binmore and Swierzbinski (2000) agree with the general results found by Ausubel and Cramton (2002); in addition, they examine empirical studies that compare the performance of discriminatory and uniform multi-unit auctions. These studies involve both "natural experiments" in which market operators have changed a discriminatory to a uniform auction or vice versa, and "laboratory experiments" in which subjects, typically students, participate in auctions designed by an experimenter. The former category of studies include a government's foreign exchange auction in Zambia, and treasury bill auctions in Mexico and the U.S.A. The authors find that empirical evidence on the question whether discriminatory or uniform price auctions are preferable is mixed. Some studies conclude that switching from a discriminatory to a uniform auction can increase the seller's revenue. Other studies suggest the reverse.

Klemperer (2002) argues that uniform price auctions facilitates collusion among bidders, because deviation from a collusive agreement is severely punished. He argues that this settlement rule is inappropriate especially for frequently repeated auctions like those for electricity.

Given the mixed results that theoretical considerations and empirical observations lead to, the question of determining the better pricing rule might be answered by simulations from agent-based electricity models.

Different research papers about AB electricity market simulations report on comparisons between pay-as-bid and uniform pricing. Results from most of these studies, e.g. Bakirtzis and Tellidou (2006), Bin et al. (2004), Cincotti et al. (2006), Xiong et al. (2004), or Weidlich and Veit (2006), allow the conclusion that the pay-as-bid mechanism leads agents to bid at higher prices, but overall average costs of electricity procurement are still lower under this mechanism than under uniform pricing. Only the findings presented by Bower and Bunn (2001) seem to confirm the theoretical considerations given by Kahn et al. (2001), i.e. that uniform price auctions lead to lower overall procurement costs.

6.2.2 Results

Simulation runs have been carried out in order to contribute to the discussion of whether uniform pricing or discriminatory pay-as-bid pricing leads to better market outcomes, i.e. to lower costs for minute reserve procurement. The two simulation scenarios UNI (uniform pricing) and PAB (pay-as bid pricing) are defined, and resulting market prices compared.

Based on the results found from other agent-based simulation models, the following hypothesis is formulated:

Hypothesis 3. *The pricing rule on the balancing power market has an effect on resulting prices for minute reserve, and market prices are higher under uniform pricing:* $P_k^{BPM,UNI} > P_k^{BPM,PAB} \ \forall k$

The day-ahead electricity market is much more important, in practice, than the balancing power market, and significantly higher volumes are traded on the former. Consequently, it can be assumed that changing prices on the balancing power market only influence day-ahead energy prices to a minor, negligible extent. This is accentuated in the second hypothesis for this simulation scenario.

Hypothesis 4. *The pricing rule on the balancing power market has no side-effect on day-ahead market prices:* $P_h^{DAM,UNI} = P^{DAM,PAB}{}_h \ \forall h$

Figure 6.4 shows a comparison of minute reserve prices for both settlement rules applied in the agent-based electricity sector model (with Q-learning). The outcome from these run seems to confirm the majority of agent-based simulation outcomes that are described in the literature. Resulting market prices for minute reserve are higher in the uniform pricing case than under pay-as-bid pricing. From Fig. 6.4, it becomes observable that price differences between minute reserve procurement auctions with uniform pricing and with pay-as-bid pricing are notable especially in the bidding blocks with high system load, with absolute differences around 14–16 EUR/MW (note that prices in this plot rank from 0 to 100 EUR/MW). The yearly average of minute reserve prices is 30.73 EUR/MW with pay-as-bid pricing, and 36.72 EUR/MW (+19.5%) in the uniform pricing case (for simulations with Q-learning).

The application of the paired-*t* confidence interval method at a 90% level confirms that resulting prices under uniform pricing are higher than under discriminatory pricing in a statistically significant manner, with a few exceptions.

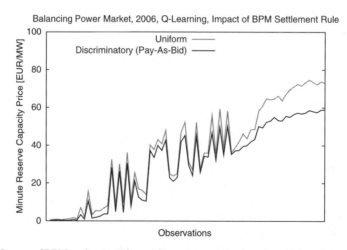

Fig. 6.4 Impact of BPM settlement rule on minute reserve capacity prices (Q-learning simulations)

6.2 Settlement Rules on the Balancing Power Market

If Erev and Roth reinforcement learning is applied, results are less definite. Minute reserve pricing under both settlement rules and under both applied variants of the learning algorithm vary strongly between runs. The variance of prices between runs is higher under uniform pricing. While average prices are also higher under uniform pricing for both learning models (+15.3% in the case of original Erev and Roth learning and +28.9% for simulations with the modified Erev and Roth algorithm), these differences cannot be regarded as statistically significant, using the 90% paired-t confidence intervals, in many of the 72 observations, due to the high variance in prices between different runs. Consequently, results from Q-learning runs are only weakly confirmed by additional Erev and Roth learning simulations. The confidence intervals for all three learning models are presented in Appendix C.2.

When comparing prices on the day-ahead market for the tested two settings with different settlement rules applied on the balancing power market, only very small differences can be observed (Fig. 6.5). Yearly average prices are less than 1% higher in the uniform pricing case than with pay-as-bid (Q-learning simulations). Similar findings come out of additional simulations with Erev and Roth reinforcement learning. Difference in day-ahead prices for the two tested scenarios are not statistically significant under all three applied learning models.

The comparison of day-ahead electricity pricing for the two pricing rule allows the conclusion that opportunity costs resulting from the balancing power market hardly affect day-ahead prices, while, vice versa, balancing power prices are strongly influenced by resulting day-ahead prices. This observation is consistent with experts' assessments about (wholesale) market interrelations in the German electricity sector.

In summary, Hypothesis 3 is only confirmed by simulations with Q-learning, but not by simulations with the Erev and Roth reinforcement learning algorithm.

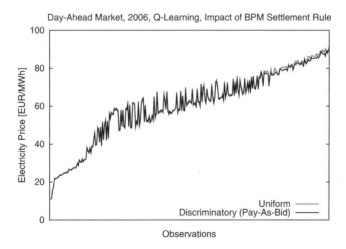

Fig. 6.5 Impact of BPM settlement rule on day-ahead electricity prices (Q-learning simulations)

Hypothesis 4, however, is not rejected by any of the three simulations with varying learning algorithms, so there is strong evidence that the settlement rule and the resulting prices on the balancing power market prices do not significantly influence day-ahead market prices.

Several alternative market rules for balancing power markets have been proposed in the literature. Schummer and Vohra (2003) for example present an incentive-compatible mechanism for balancing power markets which requires the procurer to know the distribution of demand, that is of minute reserve that actually needs to be deployed. Oren (2001) compares several design options for simultaneous auctions in which four balancing power qualities[1] are traded at the same time. Oren and Sioshansi (2004) propose to integrate balancing power procurement into the day-ahead energy auction, and to let capacity payments to undeployed reserves be based on the generator's implied opportunity cost. The mechanism that Chao and Wilson (2002) advocate for bases the decision which generators are considered for holding capacity in reserve only on the capacity price, and lets the price paid for energy delivery be defined by the spot price on the real-time market.

The presented approaches are interesting options of efficient balancing power procurement. Several changes to the implementation of the electricity sector simulation model would be necessary to analyze which of these options is most suitable for ensuring efficient market operation in the case of balancing power procurement in Germany. Carrying out the respective simulation runs is suggested for future research with the model developed within this work.

6.3 Increasing Supply Side Competition Through Divestiture

6.3.1 Motivation

In her 16th Main Report, the German Monopolies Commission[2] "regards supervision of competition in the wholesale electricity and regulated energy markets [=balancing power markets] as inadequate. The need for special supervision of competition on these markets is due to their particular vulnerability to supply strategies by generating companies that have sufficient market power to influence prices". (Monopolies Commission, 2006).

Generators with market power are usually understood as firms who can set prices above marginal generation costs while still making positive sales (cp. for example Rassenti et al., 2003).

In the German electricity sector, the four largest generating companies taken together own 70–80% of the total German generating capacity. In public debate,

---

[1] While German TSOs and the UCTE only distinguish th ree balancing qualities of primary, secondary and tertiary (minute) reserve, the California Independent System Operator procures four types of reserves.

[2] The Monopolies Commission is an advisory board on competition policy and regulation questions in energy markets and other economic sectors in Germany.

6.3 Increasing Supply Side Competition Through Divestiture

it is often presumed that these companies have market power, and that they contribute to driving prices away from competitive levels. Under the assumption that the oligopolistic structure of the German electricity sector leads to excessive wholesale prices, it would be an adequate policy measure to increase supply side competition through forcing the dominant companies to divest parts of their generation assets. Similar measures have been carried out in the electricity industry of other regions (e.g. England and Wales, cp. Day & Bunn, 2001).

The question to what extent the four large generating companies actually dispose of market power in the German electricity markets should not be discussed here. For the following analysis, it is "tacitly" assumed that they have some potential exert market power, and it is tested whether divestiture measures can mitigate this potential to some extent. If results show that plant divestiture has no influence on market prices, this may either invalidate the assumption, or it may indicate that the tested divestiture options are not appropriate to mitigate the players' market power exertion.

6.3.2 Results

In order to assess the impact of plant divestiture in the German electricity industry, six divestiture (DIV) scenarios have been defined and run with the agent-based electricity sector simulation model. Results are compared to the corresponding scenario without divestiture (NODIV). The analysis has been guided by the following hypothesis:

Hypothesis 5. *If the four largest generator agents have to divest part of their generating capacity, market prices on the day-ahead and balancing power market decrease as compared to the same scenario without plant divestiture:*
$P^{DAM,DIV}_h < P^{DAM,NODIV}_h \; \forall h$ and $P^{BPM,DIV}_k < P^{BPM,NODIV}_k \; \forall k$

The six tested divestiture scenarios differ in the percentage of owned capacity that the large agents have to divest, and in the number of newly introduced agents that take over the divested capacity. In all divestiture scenarios, it is assumed that the largest four agents defined in the model, that is Gen1, Gen2, Gen3 and Gen4,[3] representing the four dominant players of the real-world industry, have to divest a given fraction – a quarter, a third, or half – of installed capacity of all their power plant types. Either four or eight new agents are introduced into the market, that each take over the divested capacity of one of the four large firms. In the case of four agents, they then each own the whole divested capacity of one of the former large firms; in the case of eight new agents, two agents share the divested capacity of one large firm. Consequently, each new agent has a portfolio of power plants with the same proportion of technologies (gas-fired, coal-fired, hydro, etc.) as the firm that had to divest part of its capacity. The new agents are also endowed with the share of

---

[3] See Appendix B.2 for agents' characteristics.

emission allowances associated to the divested power plants. The nomenclature of the six defined scenarios are summarized in Table 6.1.

If market power exertion plays a major role in price formation on the day-ahead and/or balancing power market, it might be assumed that the stronger the large players are disintegrated, the stronger prices decrease as a result of more competitive agent bidding. This conjecture, which is to be tested with the help of the six divestiture scenarios, is formulated in two further hypotheses:

Hypothesis 6. *When the large players have to divest a bigger share of their capacity, prices on both markets will decrease stronger than if the share of divested plants is smaller:*
$$P_h^{DAM,DIVX,0.5} < P_h^{DAM,DIVX,0.33} < P_h^{DAM,DIVX,0.25} \; \forall h \text{ and}$$
$$P_k^{BPM,DIVX,0.5} < P_k^{BPM,DIVX,0.33} < P_k^{BPM,DIVX,0.25} \; \forall k$$

Hypothesis 7. *When more new agents enter into the market with divestiture, prices on both markets will decrease stronger:*
$$P_h^{DAM,DIV8,X} < P_h^{DAM,DIV4,X} \; \forall h \text{ and } P_k^{BPM,DIV8,X} < P_k^{BPM,DIV4,X} \; \forall k$$

Average monthly and yearly prices resulting from the tested scenarios are depicted in Fig. 6.6. The general assumption that divestiture leads to falling market prices (Hypothesis 5) is confirmed by the results depicted in this diagram. Yearly average prices on the day-ahead market are reduced by 6.1%, 7.4% and 8.4% for the

Table 6.1 Divestiture scenarios

Scenario name	Fraction of divested capacity	Number of new agents
DIV4,0.25	0.25	4
DIV4,0.33	0.33	4
DIV4,0.50	0.50	4
DIV8,0.25	0.25	8
DIV8,0.33	0.33	8
DIV8,0.50	0.50	8

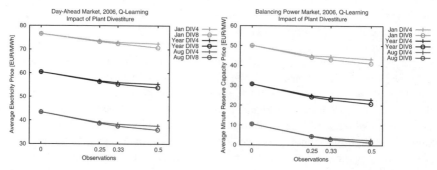

Fig. 6.6 Monthly/yearly average prices on the day-ahead (*left*) and balancing power market (*right*) for different divestiture scenarios

6.3 Increasing Supply Side Competition Through Divestiture

scenarios with four new agents, DIV4,0.25, DIV4,0.33 and DIV4,0.50, and by 6.7%, 8.5% and 10.9% for the scenarios with eight new agents, DIV8,0.25, DIV8,0.33 and DIV8,0.50, all compared to corresponding prices in the scenario without divestiture. On the balancing power market, the decline in prices is 19.1%, 22.3% and 25.9% for the scenarios DIV4,0.25, DIV4,0.33 and DIV4,0.50, and 21.0%, 25.6% and 32.3% for scenarios DIV8,0.25, DIV8,0.33 and DIV8,0.50, compared to the reference scenario (for Q-learning simulations).

Market prices in all divestiture scenarios are lower than prices in the case without divestiture in a statistically significant way, according to results from the paired-t confidence interval approach (at a 90% confidence level, for Q-learning simulations; see Appendix C.3 for an overview of the confidence intervals for all learning models). The very few (<10) exceptions among the 288 observations for each setting on the day-ahead market, in which differences in prices from the scenario with and without divestiture are not statistically significant, are the hours of highest system load. This can also be followed from Fig. 6.7, which plots resulting prices for all observations on the day-ahead market: in the last observations of very high load, prices do not decrease as a consequence of divestiture. This is a surprising result, because standard economics usually assume that market power can be exerted especially in situations of tight supply. Market power mitigation measures, such as the proposed divestiture procedures, should especially lower prices in these hours of peak demand and, consequently, weak competition.

On the balancing power market, prices are statistically significantly lower in simulations with divestiture than in those without divestiture for all observations and all scenarios. The market prices for the DIV4,0.5 and DIV8,0.5 scenario over all observations are depicted in Fig. 6.8.

If the different divestiture scenarios are compared in Fig. 6.6, it can be seen that the intuition formulated in Hypothesis 6 is not rejected by the results. Average prices

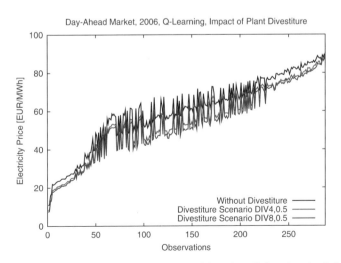

Fig. 6.7 Impact of plant divestiture on day-ahead electricity prices (Q-learning simulations)

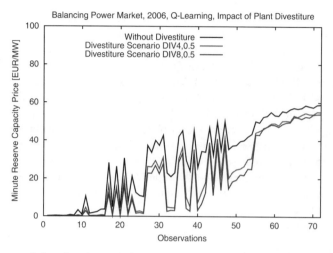

Fig. 6.8 Impact of plant divestiture on minute reserve capacity prices (Q-learning simulations)

in those scenarios with higher amounts of divested capacity are lower than in the scenarios with lower divestiture amounts. However, these differences in prices are only statistically significant for some observations. Consequently, Hypothesis 6 receives only very weak support by the simulation results.

The difference in resulting prices for divestiture scenarios with four and with eight new agents that enter into the market is only small, especially if a low share of capacity from the large generators is divested. The difference in resulting prices in the case of four and eight new players becomes more important when the share of divested capacity rises to 50%. For all scenarios, differences in prices of scenarios with four new agents are only statistically significant for some observations, so Hypothesis 7, too, receives only very weak support.

Simulations applying Erev and Roth reinforcement learning generally confirm the Q-learning results presented here. Price reductions on the day-ahead and balancing power markets as a consequence of divestiture are summarized as follows: original Erev and Roth RL simulations: 5.0%, 5.8%, 7.1%, 5.1%, 6.8% and 9.4% for the scenarios DIV4,0.25, DIV4,0.33, DIV4,0.50, DIV8,0.25, DIV8,0.33 and DIV8,0.50 on the day-ahead market and 22.8%, 21.0%, 24.3%, 20.5%, 24.5% and 34.0% for the same scenarios on the balancing power market; for simulations with the modified variant of Erev and Roth RL, price reductions in the divestiture scenarios are 5.7%, 6.8%, 7.6%, 5.7%, 7.6% and 9.7% on the day-ahead market and 14.6%, 14.1%, 17.5%, 11.0%, 12.4% and 20.1% on the balancing power market, as compared to the reference scenario.

The six divestiture options considered here all share the characteristic that portfolios of plants from the largest players had to be divested, and that the new players took over these whole portfolios. Another divestiture option would be that players are forced to divest specific types of power plants only, for example all peak load units (usually gas and oil fired plants). With this option, all agents would have less possibilities to optimize trading strategies over a diversified portfolio of different types of plants, which could additionally bring down market prices. Simulating the respective divestiture scenarios is suggested for future work.

6.4 Policy Implications and Summary

In this section, three scenarios of varying market structure or market rules have been presented and results from respective simulation runs were evaluated. The scenarios analyzed in this chapter correspond to possible policy measures that are discussed or that could be implemented in the German electricity sector. The impact of these measures on electricity market prices, as observed from the simulation results, are now compared and discussed, so that conclusions can be drawn for policy making.

From the scenario presented in Sect. 6.1, it could be seen that procured balancing power capacities have a considerable influence on resulting prices, both on the day-ahead and on the balancing power market. In current practice in Germany, the amount of minute reserve energy actually deployed for frequency regulation is only around 2% of the procured capacity for this balancing power type (Bundesnetzagentur, 2006). This illustrates that the transmission system operators include substantial buffers when calculating the balancing power capacities to be tendered. The amounts of balancing power of different types procured by the TSOs will probably increase in the future, if more electricity from fluctuating renewable energy sources, such as wind energy, are deployed. As a consequence, balancing power remains important for a functioning electricity system, and the markets at which the reserves are procured have go be carefully designed. As the results presented in Sect. 6.2 suggest, a shift from the pay-as-bid towards a uniform price settlement scheme is not likely to produce a decrease in balancing power market prices. In contrast, as the quantities procured at the balancing power market are subtracted from other wholesale power markets and lead to increasing prices there, integrating balancing power procurement into the day-ahead or real-time markets seems to be a promising option to bring about price decreases. The balancing power market design proposed by Oren and Sioshansi (2004), for example, is a good suggestion that should also be considered for the German power sector.

Independent of the actual design of balancing power markets, another policy implication arises from the aforementioned considerations. In Germany, the four TSOs are each part of holdings that also integrate generating companies, and these vertically integrated generating companies are the four largest in the national market. Obviously, generators profit from high prices on both power markets. If these belong to the same holdings which also operate the balancing power markets, they have no incentive to care for installing rules and structures that ensure competitive and efficient wholesale power markets, unless they are forced to do so by a regulatory intervention. As these vertically integrated trusts have no interest in low prices, and as the results presented in this chapter suggest that the design of the balancing power market can actually have an effect on prices, it follows that the balancing power markets should be operated by an independent party that does not generate and sell electricity. This postulation, i.e. the call for an enhanced unbundling in the power sector, is also one important point in the current energy-related discussion at the European level.

Compared to the influence of tendered minute reserve quantities, the impact of the simulated divestiture options as presented in Sect. 6.3 is surprisingly weak. The described measures only result in small reductions of resulting market prices. Given that a plant divestiture of the dimension implied by the six tested scenarios are presumably very difficult to enforce from a political perspective, the resulting effects on power market prices may not justify such drastic changes in the market structure. It seems that the ratio of supply capacity in relation to system's total load levels has a stronger influence on prices than the ownership structure. This is confirmed by the finding of the simulations presented in Sect. 6.3, which is that divestiture does not drive down prices in the times of very high system load. It seems, thus, very important to care for an environment in which enough generating capacity is built, because only sufficient capacities can prevent price spikes resulting from tight supply situations in times of very high demand. This draws a bow to the third market considered in the agent-based electricity sector model, i.e. the exchange for CO_2 emission allowances. The unclear allocation of allowances for future trading periods (beyond 2012) has been states as one important barrier to new investments, which are needed for stable prices, especially in the face of an aging power plant portfolio in which many power plants reach the end of their technical lifetime, soon. The uncertainties arising from unclear allowance allocation procedures beyond 2012 should, thus, be resolved as quickly as possible.

Chapter 7
Conclusion and Outlook

In this thesis, an agent-based electricity market simulation model has been developed and applied to several research questions. The model comprises a day-ahead market for hourly contracts of electricity delivery, a balancing power market at which positive minute reserve capacities are procured by transmission system operators, and an exchange for CO_2 emission allowances. Market participants are modeled as software agents who have learning capabilities, represented through reinforcement learning algorithms. The model has been run with data input from the German electricity sector (UCTE system's total load data and power plant portfolios that roughly correspond to those of the main players in the German power markets), and the prices resulting from the dynamic interaction of agents in the modeled markets are compared to prices observed at the corresponding real-world markets in Germany. The developed agent-based simulation model delivers realistic daily and seasonal courses of prices on the day-ahead electricity market and on the balancing power market. It can therefore be used for methodologically supporting questions of how to best *engineer* markets in the electricity sector. Some examples for this procedure have been calculated in this thesis, and conclusions for policy advice have been drawn.

The main contributions that have been achieved through the present work are summarized in Sect. 7.1. Finally, some suggestions for future work in the field analyzed with the agent-based simulation model are formulated in Sect. 7.2.

7.1 Original Contribution

The contribution of the present work is twofold: on the one hand, the developed simulation model has helped to gain insights into the dynamics of wholesale power trading under different market structures and market rules, with an application to the case of Germany. On the other hand, the development process has added to the advancement of the agent-based electricity modeling methodology itself.

From the methodological perspective, the thesis has started with a detailed and precise survey of the state-of-the-art of agent-based simulation models, applied to electricity market research. The findings from this survey led to the formulation of the current strengths and weaknesses of the methodology, and pointed out the most important fields into which further research efforts should be directed. The publication of a longer version of this survey in a reputable journal for energy-related research (Weidlich & Veit, 2008c) illustrates that such a structuring of the fast growing and still unclear research field was a very helpful scientific contribution.

One important weakness of the majority of reviewed papers that was found from this detailed analysis was the lack of proper empirical validation procedures for agent-based (electricity) models. In the current work, a two-stage validation procedure has been proposed and applied to the model development process. At the first stage, which can also be considered as the *micro*-validation, the choice of the right representation for the agents' strategic bidding behavior was in the focus of investigation. As the overall model contains a variety of parameters that have to be set to the right values, the model of agent adaptation – here, the reinforcement learning algorithms – should be analyzed in more detail within a simplified model, in order to make sure that they represent the agents' behavior in an appropriate way. This *micro*-validation procedure has been reported in Chap. 4.

The next step, i.e. the *macro*-validation of the model, was based on a graphical inspection of simulation outcomes, which were compared to real-world data, and on the statistical *confidence interval approach* proposed by Law (2007). In summary, a good fit of simulated market prices with those observed in real-world markets was found. This set the ground for the successful application of the developed model to the research questions investigated within this work. What could also be concluded from the results of the validation procedure is that the application of Q-learning leads to better simulation results, both in terms of similarity of outcomes to real-world data and in terms of variance of results between runs. The simulated prices of different simulation runs applying different random number seeds were very similar to each other for simulations with Q-learning, while the variance of results between runs with the Erev and Roth reinforcement learning algorithm was quite large. Simulations applying the Erev and Roth learning model were found to be only useful for giving support to the outcomes of Q-learning simulations. This is a further methodological proposition of this work: it postulates that simulation results that come out of simulations applying one learning model should be confirmed by at least one additional series of runs applying an alternative learning algorithm variant, in order to make sure that conclusions are based on a wide basis of evidence. Throughout the simulations carried out in this work, results from Erev and Roth simulations confirmed the tendencies found from Q-learning simulations.

The model developed within this work provides insights into the aggregate system characteristics resulting from dynamic strategic behavior in oligopolies under consideration of learning from daily interaction. It might help a regulator to prevent market power exertion through introducing suitable market designs or enforcing changes in the market structure. From the results presented here, some general characteristics can be encountered:

7.1 Original Contribution

- Resulting simulated market prices on the day-ahead and on the balancing power market are strongly influenced by the demand and supply ratio. The level of system's total load determines electricity price levels and also the prices for CO_2 emission allowances to a great extent. In addition, if the amount of demand on one power market, or the available capacities on the supply side of a market are altered through measures like the variation of tendered minute reserve capacities (as described in Sect. 6.1), the impact on prices is considerable.
- The ownership structure of the generating capacities is less influential on market prices than the ratio of demand and supply. This has been observed through the divestiture scenarios presented in Sect. 6.3. Although divestiture leads to decreasing market prices on both the day-ahead and the balancing power market, as it was expected, the price decrease is rather weak as compared to a variation of demand quantities.
- Opportunity costs play an important role for price formation on the balancing power market, because the day-ahead electricity prices are an important reference for the agents' daily minute reserve bidding prices. Through these opportunity costs, both markets are strongly interrelated. Similarly, prices for CO_2 emission allowances influence electricity prices in a considerable way, and CO_2 prices depend on the emission situation of the simulated month. Consequently, the implication of policy measures must be evaluated on all three markets simultaneously.

Some policy implications were derived from the simulation results gained from the agent-based electricity sector model. Among these are the suggestion to integrate balancing power procurement auctions into the day-ahead electricity market, or to link it to real-time trading, because the tendered balancing power capacities actually have an influence on resulting market prices, and the required quantities might still rise as a consequence of an increased deployment of fluctuating renewable energy sources. With an integration of both markets, the full available capacity can be bid on both products simultaneously, potentially leading to lower prices both for electricity delivery and for holding balancing capacity in reserve.

Given the strategic potential that balancing power markets have, it is furthermore advocated to fully unbundle operation of the transmission systems – which also entails operating the balancing power markets – from power generation, so that balancing market operators have no incentives to take measures that lead to increasing electricity prices.

The presented simulation model was motivated by the postulation formulated similarly by Roth (2002), Weinhardt et al. (2003), and Marks (2006), i.e. that markets should be designed using engineering tools such as experimentation and computation. Agent-based simulation offers one valuable tool within the process of a structured approach to *engineering* complex marketplaces such as electricity markets, and the present work has added to the development and application of this methodology to questions of shaping markets in the German power sector.

Good market design is crucial for electricity markets, and many problems have to be solved if efficient and competitive market outcomes are to be achieved (see e.g. Cramton, 2003 for a comprehensible overview). The agent-based electricity sector model developed within the present work is one suitable element that helps to answer a variety of these questions relevant to the research field.

7.2 Outlook on Possible Future Work

The simulation model developed within this work may be enhanced in several way. First, as the German electricity system is highly intermeshed with those systems of the neighboring countries, the national perspective followed within this work is a considerable simplification. In a future development of the simulation model, a European view on wholesale power trading should be adopted, and the electricity systems of the most important countries around Germany should be modeled explicitly.

A European power trading model automatically implies that transmission becomes a more important topic. The current version of the model neglects transmission constraints. For the research questions investigated here, and with the national focus adopted, disregarding the physical transmission system is an acceptable simplification, because transportation capacities are rarely binding for wholesale power trading within Germany. As soon as a European perspective is considered, or if possible future developments of the physical power system are tested, a more realistic transmission system representation should be embraced into the model. A model development into this direction has already been started, and first tests have shown that the inclusion of transmission constraints has an influence on regional prices in some extreme cases of renewable and conventional energy availability.

While the focus of investigation in this thesis was placed on overall market prices that result from the strategic interaction of players in the electricity markets, it would also be interesting to analyze the micro level of the agents' trading strategies in more detail. With the developed simulation model, it can easily be analyzed how varying market structures or policy measures influence trading success and profits of single agents. If, for example, regulatory intervention aims at promoting specific types of generating technologies the effectiveness of different conceivable actions into this direction could be tested through structured simulation runs with the developed model, with a subsequent analysis of resulting bidding strategies for the different power plant types. It is planned to enhance research work into this direction.

Some other aspects that have not yet been considered in the simulation model developed here, but which can be accounted for in agent-based electricity research, are for example the implications of an actively bidding demand side, or the sustainability of collusive strategies by supply side agents. Also, a comparison of a vertically integrated electricity industry structure with a disintegrated structure would be helpful for quantifying the value of unbundling activities. Lastly, some more levels of electricity trading, such as forward and real-time trading or additional reserve qualities on the balancing power market may help to make the current model a more realistic and valid representation of the real-world electricity trading system.

The possibilities offered by the rich methodology of agent-based simulation for electricity-related research are not yet fully utilized, so a number of further research questions can be treated with future developments of the model presented within this work.

Appendix A
Learning Model Testing Scenarios

A.1 Definition of the Two-Dimensional Action Domain and Spillover of Reinforcement

The set of actions that an agent can choose from consists of the two dimensions *bid price* and *bid quantity*. The resulting action domain is depicted in Fig. A.1, with rows representing possible price levels and columns representing quantity levels (relative to available capacity).

The computational implementation of the learning algorithm handles action domains as one-dimensional arrays of actions. The *learner*, that is the actual learning algorithm applied, is only concerned with evaluating the trading success associated with the last chosen action. It knows the (integer) index of this chosen action, but does not need to know what characteristics the action has. The class `ActionDomain` arranges the mapping between index and content of an action and is called upon by the adaptive agent. This generic implementation allows to represent a variety of possible action domains. In the case considered here, the two-dimensional action domain has to be collapsed into a single-dimensional form. The source code of this procedure is summarized in the following code cutout.

```
double[][] actions;

public ActionDomain (int dim1, double min1, double max1,
        int dim2, double min2, double max2) {
    numActions = dim1 * dim2;
    int index = 0;
    actions = new double[numActions][2];
    for (int i = 0; i < numActions1; i++) {
        for (int j = 0; j < numActions2; j++) {
            actions[index][0] = i * (max1 - min1) / (numActions1 - 1);
            actions[index][1] = j * (max2 - min1) / (numActions2 - 1);
            index++;
        }
    }
}
```

0		1		2		3		4		5	
	0,0		0,20		0,40		0,60		0,80		0,100
6	5,0	7	5,20	8	5,40	9	5,60	10	5,80	11	5,100
12	10,0	13	10,20	14	10,40	15	10,60	16	10,80	17	10,100
18	15,0	19	15,20	20	15,40	21	15,60	22	15,80	23	15,100
24	20,0	25	20,20	26	20,40	27	20,60	28	20,80	29	20,100
30	25,0	31	25,20	32	25,40	33	25,60	34	25,80	35	25,100
36	30,0	37	30,20	38	30,40	39	30,60	40	30,80	41	30,100
42	35,0	43	35,20	44	35,40	45	35,60	46	35,80	47	35,100
48	40,0	49	40,20	50	40,40	51	40,60	52	40,80	53	40,100
54	45,0	55	45,20	56	45,40	57	45,60	58	45,80	59	45,100
60	50,0	61	50,20	62	50,40	63	50,60	64	50,80	65	50,100
66	55,0	67	55,20	68	55,40	69	55,60	70	55,80	71	55,100
72	60,0	73	60,20	74	60,40	75	60,60	76	60,80	77	60,100
78	65,0	79	65,20	80	65,40	81	65,60	82	65,80	83	65,100
84	70,0	85	70,20	86	70,40	87	70,60	88	70,80	89	70,100
90	75,0	91	75,20	92	75,40	93	75,60	94	75,80	95	75,100
96	80,0	97	80,20	98	80,40	99	80,60	100	80,80	101	80,100
102	85,0	103	85,20	104	85,40	105	85,60	106	85,80	107	85,100
108	90,0	109	90,20	110	90,40	111	90,60	112	90,80	113	90,100
114	95,0	115	95,20	116	95,40	117	95,60	118	95,80	119	95,100
120	100,0	121	100,20	122	100,40	123	100,60	124	100,80	125	100,100

Fig. A.1 Action domain at the day-ahead electricity market with action indices

In a two-dimensional action space, similar actions do not always have neighboring indices, and actions with neighboring indices are not always similar. One action has a maximum of four similar neighboring actions to which some part of the reinforcement is to be spilled over in a similar manner as formulated in (4.2). As depicted in Fig. A.1, for example, the action $(25,40)$ with index 32, corresponding to a bid of 40% of available capacity at a price of 25 EUR/MWh, is similar to a bid with the same quantity at the next higher or next lower possible price (actions with indices 26 and 38). It is also similar to a bid at the same price for a next higher or next lower possible quantity (actions with indices 31 and 33). Formula 4.2 has to be redefined for two-dimensional action domains in the following manner:

$$q_{ij}(t+1) = \begin{cases} (1-\phi)q_{ij}(t)+R(x)(1-\varepsilon) & \text{if } j=k \\ (1-\phi)q_{ij}(t)+R(x)\frac{\varepsilon}{4} & \text{if } j=k\pm dim2 \\ & \vee (j=k+1 \wedge (k+1) \bmod dim2 \neq 0) \\ & \vee (j=k-1 \wedge k \bmod dim2 \neq 0) \\ (1-\phi)q_{ij}(t) & \text{otherwise} \end{cases}$$

(A.1)

Here, *dim2* is the number of possible actions of the second dimension (i.e. six for the action domain of Fig. A.1), if the collapsing of the two dimensions to a one-dimensional array is performed in the way as described in the code example provided previously. The operator *mod* is the modulo function. Note that in contrast to (4.2), the fraction of reinforcement that is spilled over to similar actions is divided by four instead of two, as there are for similar actions.

A.2 Simulation Results for Appropriate Learning Variants

Table A.1 summarizes the simulated characteristics of those learning variants that are applied in the extensive electricity sector model. Following the descriptions of the quality criteria for learning algorithms formulated in Sect. 4.3.1, quantitative values for these criteria are given. The following information is provided:

1. The outcome of simulations with the given learning variant is expressed in terms of mean price level during the last 200 iterations of each run, averaged over 50 runs.
2. The *convergence* criterion is expressed in Coefficient of variation (CV) of prices within one simulation run, at the end of run (EOR), i.e. during the last 200 iterations; here, the average of the CVs of 50 simulation runs are reported.
3. In the *robustness* column, only the robustness against different random number seeds is reported. This is expressed in CV between runs and quantifies how much results for individual runs differ across 50 runs. As reported in Sect. 4.3.3, all algorithms are quite robust against varying initial propensities or Q-values and against the density of possible actions within a given range of prices and quantities. The variances of results for cases in which the range of possible bid prices varies are reported in Table 4.3.
4. The column named *collective rationality* summarizes the mean number of bid rounds without market clearing, i.e. the number of rounds in which the bid supply quantity was not sufficient to satisfy demand, among the last 200 iterations of a simulation run (termed "uncleared iterations").
5. As a criterion for *individual rationality*, the correlation between marginal costs and bid price is depicted in a way similar to Fig. 4.6.

Table A.1 Summarized overview of agent-based electricity market modeling approaches

Learning model	Specification	Parameter values	Price result	Convergence	Robustness	Collective rationality	Minimal rationality
			Mean price EOR	CV of prices within EOR	CV of prices between runs	Mean # uncleared iterations	Marginal costs / bid price correlation
Original Erev and Roth RL	With spillover, proportional action selection	$\phi = 0.1$, $\varepsilon = 0.2$, $q_0 = 1.0$	49.92	0.22	0.26	5.66	
Modified Erev and Roth RL	Proportional action selection	$\phi = 0.1$, $\varepsilon = 0.2$, $q_0 = 1.0$	51.10	0.00	0.30	0.00	
Modified Erev and Roth RL	Proportional action selection	$\phi = 0.1$, $\varepsilon = 0.4$, $q_0 = 1.0$	52.10	0.00	0.32	0.00	
Original Erev and Roth RL	With spillover, Softmax action selection	$\phi = 0.1$, $\varepsilon = 0.2$, $\frac{1}{\tau} = 10$, $q_0 = 1.0$	49.43	0.41	0.17	8.86	

A.2 Simulation Results for Appropriate Learning Variants

Modified Erev and Roth RL	Softmax action selection	$\phi = 0.1, \varepsilon = 0.2,$ $\frac{1}{\tau} = 12, q_0 = 1.0$	54.85	0.31	0.35	6.98
Q-learning	ε-greedy action selection	$\alpha = 0.5,$ $\gamma = 0.85, \varepsilon = 0.1$	57.85	0.40	0.07	11.46
Q-learning	Softmax action selection	$\alpha = 0.5,$ $\gamma = 0.85, \frac{1}{\tau} = 18$	49.82	0.40	0.05	8.6

Appendix B
Reference Scenario

B.1 Demand Side Data Input for the Day-Ahead Market

German system's total load for every hour of the third Wednesday of a month in the year 2006, as published by the UCTE on
http://www.ucte.org/services/onlinedatabase/consumption/

Table B.1 UCTE load for 2006 in MW

Hour	18 Jan	15 Feb	15 Mar	19 Apr	17 May	21 Jun	19 Jul	16 Aug	20 Sep	18 Oct	15 Nov	20 Dec
1	62,375	61,470	62,240	51,936	49,561	50,971	50,095	46,923	49,779	53,345	55,314	57,873
2	60,353	59,027	59,973	49,518	47,429	48,839	47,854	44,894	47,880	51,020	52,717	55,479
3	59,000	57,962	58,551	48,761	46,237	47,581	46,428	43,526	46,990	49,971	52,519	54,652
4	58,829	57,820	58,372	49,392	46,602	47,831	46,316	43,671	47,184	50,707	53,916	55,235
5	60,123	58,809	59,094	51,432	47,882	48,662	47,228	44,648	48,428	52,464	56,098	56,512
6	62,543	61,100	61,541	54,820	50,853	51,419	49,643	48,327	52,703	56,412	60,097	59,286
7	68,885	67,243	66,820	59,984	59,158	59,973	57,395	55,665	63,400	65,675	68,467	66,615
8	76,178	74,070	71,428	65,742	66,690	67,171	64,592	62,141	69,379	73,157	74,912	74,953
9	78,053	76,612	73,906	69,254	69,521	70,549	68,058	66,012	71,791	74,654	76,280	77,045
10	77,658	77,040	73,968	70,162	70,295	71,430	69,584	67,112	71,537	74,214	75,324	76,486
11	78,574	78,274	74,828	70,979	71,639	72,945	70,976	68,708	72,201	74,903	75,537	76,834
12	79,709	79,787	76,196	72,187	73,208	74,179	72,502	70,660	73,782	75,809	76,237	77,459
13	78,712	79,651	75,556	71,161	71,878	72,973	71,289	69,891	72,841	74,649	75,457	76,821
14	78,110	79,428	75,056	70,051	70,998	72,050	70,283	68,435	72,354	73,437	74,789	76,300
15	77,357	78,390	73,965	68,683	69,636	70,949	69,004	67,027	70,835	72,203	73,589	75,739
16	76,737	77,616	73,165	67,387	68,164	69,927	67,859	65,910	69,497	70,908	72,941	75,762
17	76,535	76,596	71,654	65,715	66,475	68,575	66,776	64,336	67,759	69,475	74,468	77,925
18	80,219	78,462	72,113	65,445	66,057	67,830	66,703	63,894	67,515	70,046	80,093	80,230
19	80,379	80,750	75,712	65,693	65,587	67,354	66,310	63,607	67,697	74,525	79,914	79,190
20	77,828	78,247	77,549	65,587	64,678	66,559	65,353	63,065	69,867	76,385	77,082	76,813
21	73,517	73,995	73,232	65,860	62,813	63,954	62,813	61,560	70,817	71,745	72,437	72,798
22	71,487	71,253	70,679	65,247	62,616	61,613	61,613	61,067	65,474	67,035	69,091	69,847
23	69,987	69,959	69,112	62,207	59,284	60,192	60,387	56,932	59,487	62,771	65,646	67,836
24	65,713	64,789	64,482	56,254	53,210	55,343	54,751	51,206	53,754	57,179	60,074	61,858

B.2 Supply Side Data Input: Generators and Plants

Table B.2 Generator agents and power plant characteristics of the reference scenario

Agent	Plant	Fuel	Net capacity (MW)	Variable costs (€/MWh)	No-load costs (€/MW)	Availability (€/h·MW)	MR-able[a]
Gen1	Gen1_BO	Biomass/other	1,000	20	200	1	
	Gen1_HC	Hard-coal	8,000	22	2,300	1	
	Gen1_HY	Hydro energy	1,500	1.5	0	0.7	✓
	Gen1_LIG	Lignite	1,500	4.5	3,300	1	
	Gen1_NG	Natural gas	4,000	45	100	1	✓
	Gen1_NUC	Nuclear	8,500	2	2,300	1	
	Gen1_OIL	Oil	2,000	60	200	1	✓
Gen2	Gen2_BO	Biomass/other	1,000	20	200	1	
	Gen2_HC	Hard coal	10,000	22	2,300	1	
	Gen2_HY	Hydro energy	2,500	1.5	0	0.7	✓
	Gen2_LIG	Lignite	10,000	4.5	3,300	1	
	Gen2_NG	Natural gas	4,000	45	100	1	✓
	Gen2_NUC	Nuclear	5,500	2	2,300	1	
	Gen2_OIL	Oil	500	60	200	1	✓
Gen3	Gen3_BO	Biomass/other	1,000	20	200	1	
	Gen3_HC	Hard coal	500	22	2,300	1	
	Gen3_HY	Hydro energy	3,000	1.5	0	0.7	✓
	Gen3_LIG	Lignite	8,000	4.5	3,300	1	
	Gen3_NG	Natural gas	1,000	45	100	1	✓
	Gen3_NUC	Nuclear	2,000	2	2,300	1	
Gen4	Gen4_BO	Biomass/other	1,000	20	200	1	
	Gen4_HC	Hard coal	4,000	22	2,300	1	
	Gen4_HY	Hydro energy	2,000	1.5	0	0.7	✓
	Gen4_LIG	Lignite	1,500	4.5	3,300	1	
	Gen4_NG	Natural gas	1,500	45	100	1	✓
	Gen4_NUC	Nuclear	4,500	2	2,300	1	
Gen5	Gen5_HC	Hard coal	3,000	22	2,300	1	
Gen6	Gen6_HC	Hard coal	2,500	22	2,300	1	
Gen7	Gen7_HY	Hydro energy	1,000	1.5	0	0.7	✓
Gen8	Gen8_NG	Natural gas	3,500	45	100	1	✓
Gen9	Gen9_NG	Natural gas	3,500	45	100	1	✓
Gen10	Gen10_NG	Natural gas	2,000	45	100	1	✓
Gen11	Gen11_OIL	Oil	1,500	60	200	1	✓
Gen12	Gen12_BO	Biomass/other	1,000	20	200	1	
	Gen12_OIL	Oil	1,500	60	200	1	✓
Gen13	Gen13_WIN	Wind energy	18,500	1.5	0	0.15	

[a] MR = minute reserve; a checkmark indicates that the power plant fulfills the technical requirements to deliver tertiary balancing power

B.3 Agent Characteristics on the CO_2 Market

Table B.3 Emissions scenario: generating agents

Agent	Plant	Fuel	Net capacity	Efficiency factor	Emission factor (t CO_2/MWh)	Initial allocation (kEUA)[a]
Gen1	Gen1_HC	Hard-coal	8,000	0.4	0.85	38,000
	Gen1_LIG	Lignite	1,500	0.37	1.1	14,500
	Gen1_NG	Natural gas	4,000	0.42	0.48	8,200
	Gen1_OIL	Oil	2,000	0.42	0.64	3,300
Gen2	Gen2_HC	Hard coal	10,000	0.4	0.85	48,600
	Gen2_LIG	Lignite	10,000	0.37	1.1	96,800
	Gen2_NG	Natural gas	4,000	0.42	0.48	8,200
	Gen2_OIL	Oil	500	0.42	0.64	800
Gen3	Gen3_HC	Hard coal	500	0.4	0.85	2,400
	Gen3_LIG	Lignite	8,000	0.37	1.1	77,500
	Gen3_NG	Natural gas	1,000	0.42	0.48	2,000
Gen4	Gen4_HC	Hard coal	4,000	0.4	0.85	19,400
	Gen4_LIG	Lignite	1,500	0.37	1.1	14,500
	Gen4_NG	Natural gas	1,500	0.42	0.48	3,000
Gen5	Gen5_HC	Hard coal	3,000	0.4	0.85	14,600
Gen6	Gen6_HC	Hard coal	2,500	0.4	0.85	12,000
Gen8	Gen8_NG	Natural gas	3,500	0.42	0.48	7,200
Gen9	Gen9_NG	Natural gas	3,500	0.42	0.48	7,200
Gen10	Gen10_NG	Natural gas	2,000	0.42	0.48	4,100
Gen11	Gen11_OIL	Oil	1,500	0.42	0.64	2,500
Gen12	Gen12_OIL	Oil	1,500	0.42	0.64	2,500

[a] 1,000 European emission allowances

Table B.4 Emissions scenario: other CO_2 emissions trading participants

Agent	Industry sector	Valuation (EUR/EUA)	Initial allocation (kEUA)
CO2_1	Glass	25	2,350
CO2_1	IronSteel	10	16,850
CO2_2	Lime	30	4,650
CO2_3	Paper	20	2,500
CO2_4	PulpIndustry	15	750
CO2_5	Refineries	7	12,200
CO2_6	Cement	5	11,850
CO2_7	Ceramics	17	1,250
CO2_D	Demand	35	−120,000

B.4 Simulation Results for the Reference Scenario

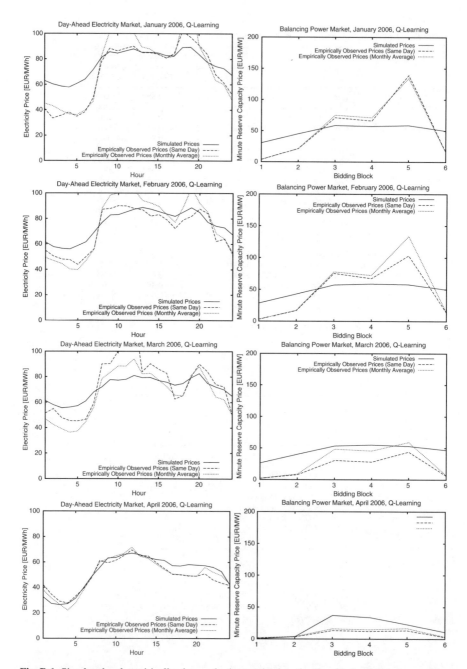

Fig. B.1 Simulated and empirically observed prices at the day-ahead and balancing power market, January–April 2006, with Q-learning, ε-greedy action selection, learning rate $\alpha = 0.5$, discount rate $\gamma = 0.9$, $\varepsilon = 0.2$

B.4 Simulation Results for the Reference Scenario

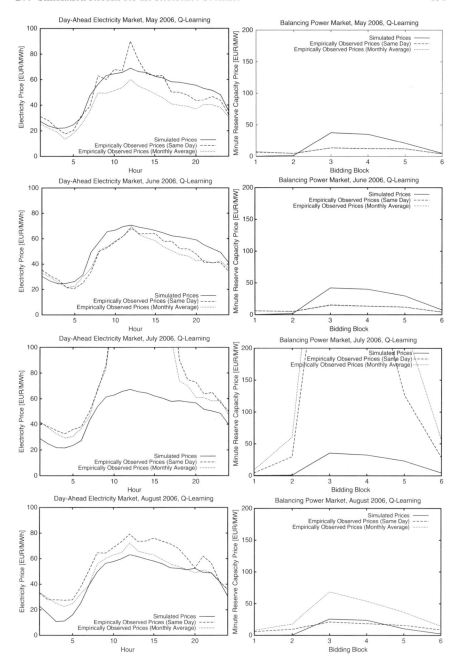

Fig. B.2 Simulated and empirically observed prices at the day-ahead and balancing power market, May–August 2006, with Q-learning, ε-greedy action selection, learning rate $\alpha = 0.5$, discount rate $\gamma = 0.9$, $\varepsilon = 0.2$

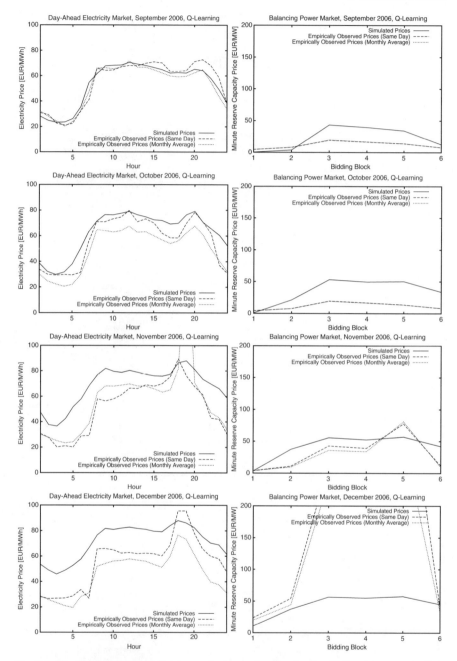

Fig. B.3 Simulated and empirically observed prices at the day-ahead and balancing power market, September–December 2006, with Q-learning, ε-greedy action selection, learning rate $\alpha = 0.5$, discount rate $\gamma = 0.9$, $\varepsilon = 0.2$

B.4 Simulation Results for the Reference Scenario

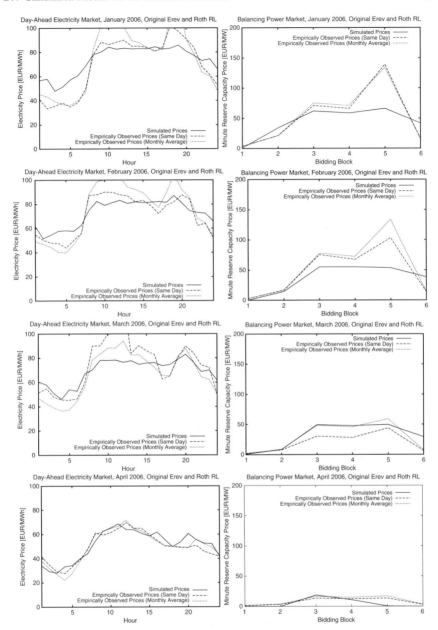

Fig. B.4 Simulated and empirically observed prices at the day-ahead and balancing power market, January–April 2006, with original Erev and Roth reinforcement learning, proportional action selection, recency $\phi = 0.1$, experimentation $\varepsilon = 0.2$

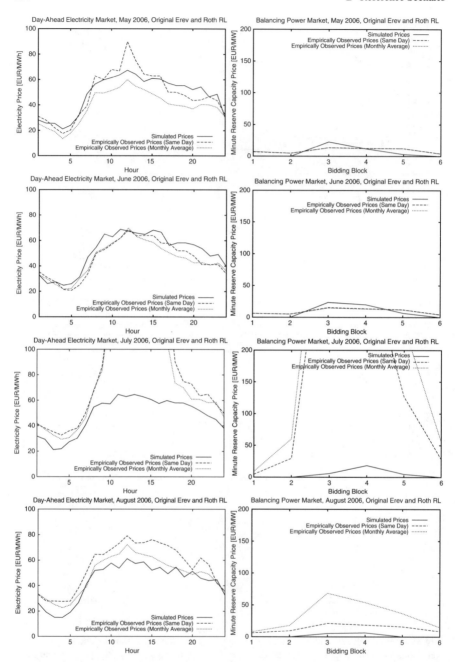

Fig. B.5 Simulated and empirically observed prices at the day-ahead and balancing power market, May–August 2006, with original Erev and Roth reinforcement learning, proportional action selection, recency $\phi = 0.1$, experimentation $\varepsilon = 0.2$

B.4 Simulation Results for the Reference Scenario

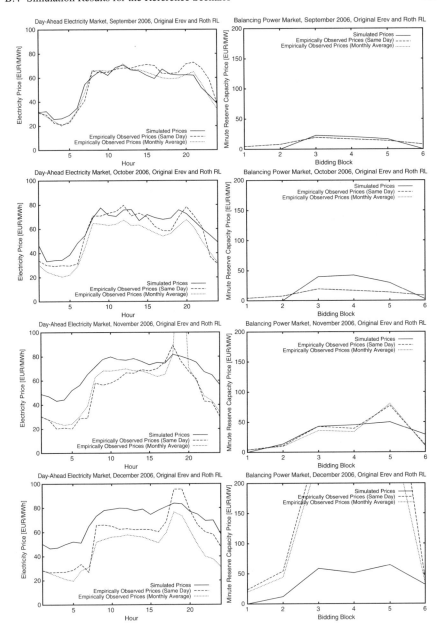

Fig. B.6 Simulated and empirically observed prices at the day-ahead and balancing power market, September–December 2006, with original Erev and Roth reinforcement learning, proportional action selection, recency $\phi = 0.1$, experimentation $\varepsilon = 0.2$

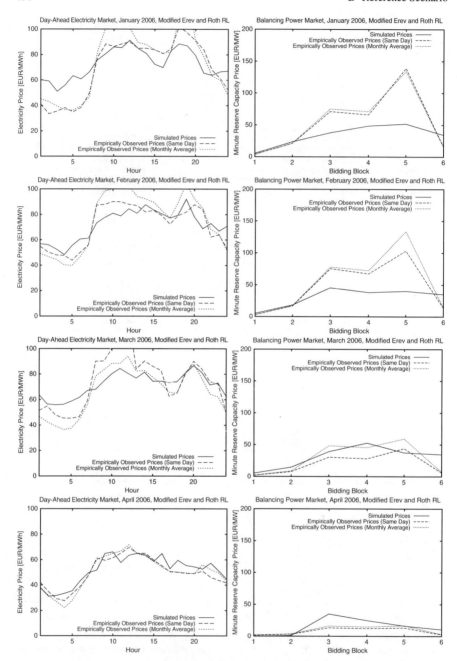

Fig. B.7 Simulated and empirically observed prices at the day-ahead and balancing power market, January–April 2006, with modified Erev and Roth reinforcement learning, proportional action selection, recency $\phi = 0.1$, experimentation $\varepsilon = 0.2$

B.4 Simulation Results for the Reference Scenario

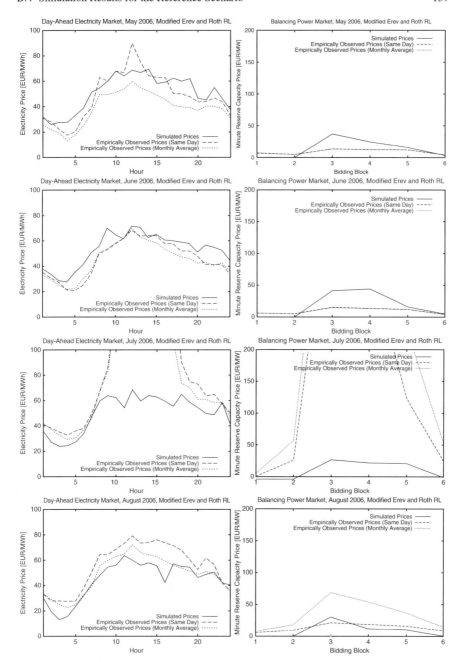

Fig. B.8 Simulated and empirically observed prices at the day-ahead and balancing power market, May–August 2006, with modified Erev and Roth reinforcement learning, proportional action selection, recency $\phi = 0.1$, experimentation $\varepsilon = 0.2$

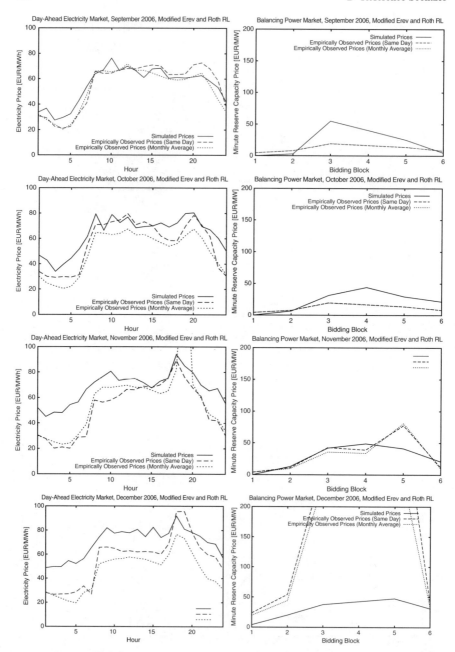

Fig. B.9 Simulated and empirically observed prices at the day-ahead and balancing power market, September–December 2006, with modified Erev and Roth reinforcement learning, proportional action selection, recency $\phi = 0.1$, experimentation $\varepsilon = 0.2$

B.5 Confidence Intervals of Sensitivity Analysis

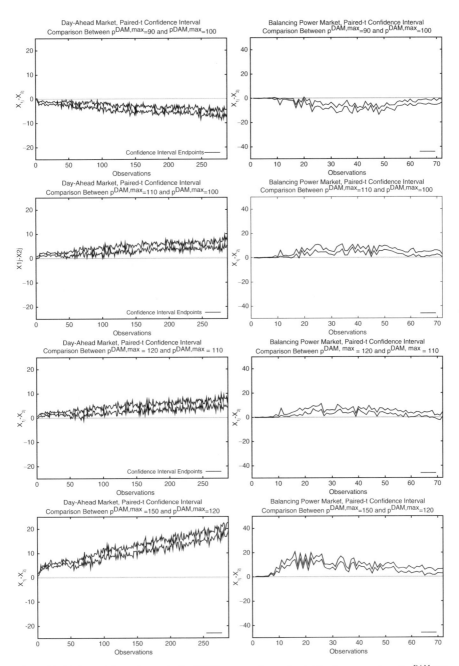

Fig. B.10 90% confidence intervals for $E(Z_j)$, comparison of neighboring scenarios of $p^{DAM,max}$ (action domains; Q-learning simulations)

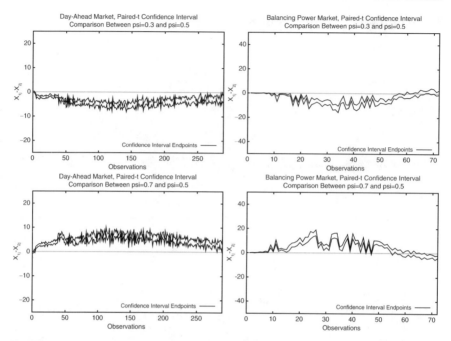

Fig. B.11 90% confidence intervals for $E(Z_j)$, comparison of different values of portfolio integration parameter ψ with reference value (Q-learning simulations)

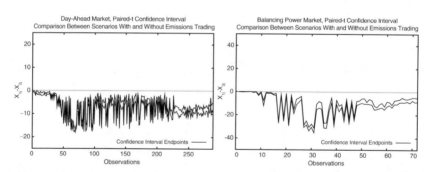

Fig. B.12 90% confidence intervals for $E(Z_j)$, comparison of scenario without emissions trading with reference scenario (Q-learning simulations)

Appendix C
Statistical Analysis of Market Scenarios

C.1 Tendered Balancing Capacity

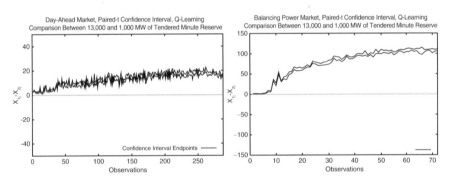

Fig. C.1 90% confidence intervals for $E(Z_j)$, comparison of lowest and highest scenarios of tendered balancing power capacity (Q-learning simulations)

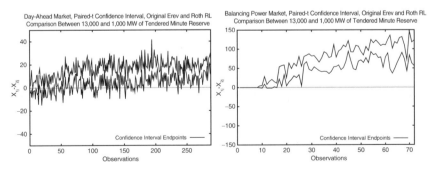

Fig. C.2 90% confidence intervals for $E(Z_j)$, comparison of lowest and highest scenarios of tendered balancing power capacity (original Erev and Roth RL simulations)

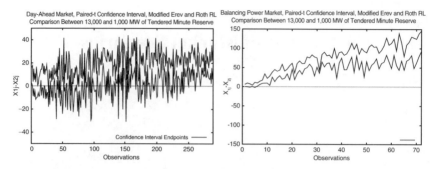

Fig. C.3 90% confidence intervals for $E(Z_j)$, comparison of lowest and highest scenarios of tendered balancing power capacity (modified Erev and Roth RL simulations)

C.2 BPM Settlement Rule

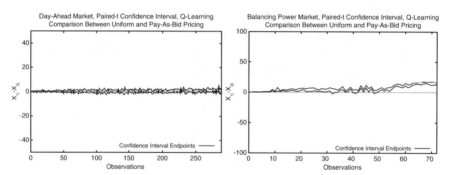

Fig. C.4 90% confidence intervals for $E(Z_j)$, comparison of pay-as-bid and uniform pricing (Q-learning simulations)

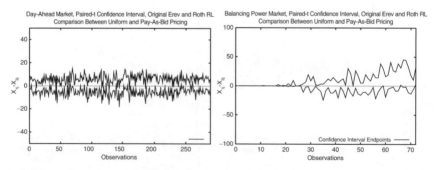

Fig. C.5 90% confidence intervals for $E(Z_j)$, comparison of pay-as-bid and uniform pricing (original Erev and Roth RL simulations)

C.3 Divestiture Scenarios

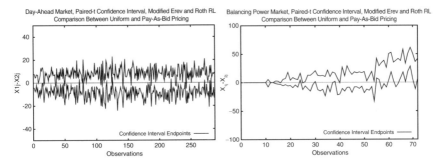

Fig. C.6 90% confidence intervals for $E(Z_j)$, comparison of pay-as-bid and uniform pricing (modified Erev and Roth RL simulations)

C.3 Divestiture Scenarios

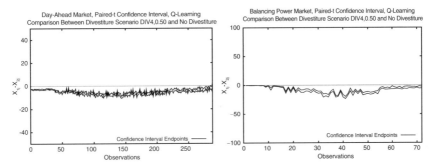

Fig. C.7 90% confidence intervals for $E(Z_j)$, comparison of DIV4,0.50 scenario with reference scenario (Q-learning simulations)

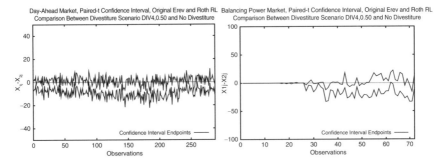

Fig. C.8 90% confidence intervals for $E(Z_j)$, comparison of DIV4,0.50 scenario with reference scenario (original Erev and Roth RL simulations)

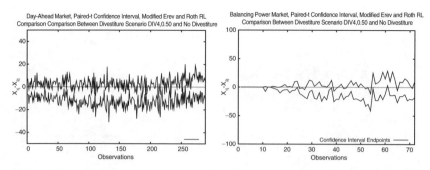

Fig. C.9 90% confidence intervals for $E(Z_j)$, comparison of DIV4,0.50 scenario with reference scenario (modified Erev and Roth RL simulations)

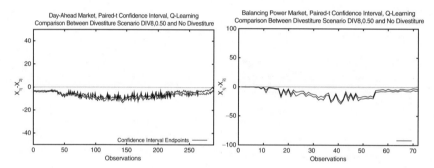

Fig. C.10 90% confidence intervals for $E(Z_j)$, comparison of DIV8,0.50 scenario with reference scenario (Q-learning simulations)

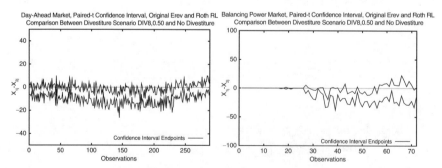

Fig. C.11 90% confidence intervals for $E(Z_j)$, comparison of DIV8,0.50 scenario with reference scenario (original Erev and Roth RL simulations)

C.3 Divestiture Scenarios

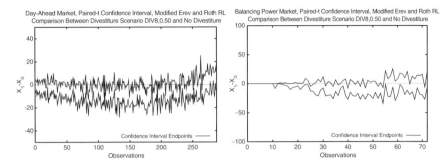

Fig. C.12 90% confidence intervals for $E(Z_j)$, comparison of DIV8,0.50 scenario with reference scenario (modified Erev and Roth RL simulations)

References

Arifovic, J., & Ledyard, J. (2004). Scaling up learning models in public good games. *Journal of Public Economic Theory*, *6*(2), 203–238.

Arthur, W. B. (1991). Designing economic agents that act like human agents: A behavioral approach to bounded rationality. *The American Economic Review*, *81*(2), 353–359.

Arthur, W. B. (1994). Inductive reasoning and bounded rationality. *American Economic Review (Papers and Proceedings)*, *84*(2), 406–411.

Atkins, K., Barrett, C. L., Homan, C. M., Marathe, A., Marathe, M. V., & Thite, S. (2004). Agent based economic analysis of deregulated electricity markets. In *6th IAEE European Conference "modelling in energy economics and policy"*. Zürich, Switzerland.

Auer, P., Cesa-Bianchi, N., Freund, Y., & Schapire, R. (2002). The nonstochastic multiarmed bandit problem. *Siam Journal of Computing*, *32*(1), 48–77.

Ausubel, L. M., & Cramton, P. (2002). *Demand reduction and inefficiency in multi-unit auctions.* Papers of Peter Cramton from University of Maryland, Department of Economics.

Axelrod, R. (1987). The evolution of strategies in the iterated prisoner's dilemma. In L. Davis (Ed.), *Genetic algorithms and simulated annealing* (pp. 32–41). San Francisco: Morgan Kaufmann.

Axelrod, R. (1997). *The complexity of cooperation: Agent-based models of competition and collaboration*. Princeton: Princeton University Press.

Axelrod, R. (2006). Advancing the art of simulation in the social sciences. In J.-P. Rennard (Ed.), *Handbook of research on nature inspired computing for economy and management*. Hersey, PA: Idea.

Axtell, R. (2000). *Why agents? on the varied motivations for agent computing in the social sciences*. The Brookings Institution, Center on Social and Economic Dynamics, Working Paper No. 17.

Axtell, R. (2005). The complexity of exchange. *The Economic Journal*, *115*, F193–F210.

Bagnall, A. J. (2004). A multi-agent model of the the UK market in electricity generation. In *Applications of learning classifier systems* (pp. 167–181). Berlin: Springer.

Bagnall, A. J., & Smith, G. (1999). An adaptive agent model for generator company bidding in the UK power pool. In *Proceedings of artificial evolution* (Vol. 1829, pp. 191–203). Springer.

Bagnall, A. J., & Smith, G. (2000). Game playing with autonomous adaptive agents in a simplified economic model of the UK market in electricity generation. In *IEEE international conference on power system technology* (pp. 891–896).

Bagnall, A. J., & Smith, G. (2005). A multi-agent model of the UK market in electricity generation. *IEEE Transactions on Evolutionary Computation*, *9*(5), 522–536.

Bagnall, A. J. (2000a). *Modelling the UK market in electricity generation with autonomous adaptive agents*. Unpublished doctoral dissertation, University of East Anglia, School of Information Systems.

References

Bagnall, A. J. (2000b). A multi-adaptive agent model of generator bidding in the UK market in electricity. In D. Whitely, D. Goldberg, E. Cantú-Paz, L. Spector, I. Parmee, & H.-G. Beyer (Eds.), *Proceedings of the genetic and evolutionary computation conference (gecco 2000)*. San Francisco: Morgan Kaufmann.

Bakirtzis, A. G., & Tellidou, A. C. (2006). Agent-based simulation of power markets under uniform and pay-as-bid pricing rules using reinforcement learning. In *Proceedings of the IEEE power systems conference and exposition psce '06* (pp. 1168–1173). Atlanta, USA.

Batten, D. (2000). *Discovering artificial economics – how agents learn and economies evolve*. Boulder: Westview.

Batten, D., & Grozev, G. (2006). NEMSIM: Finding ways to reduce greenhouse gas emissions using multi-agent electricity modelling. In P. Perez & D. Batten (Eds.), *Complex science for a complex world: Exploring human ecosystems with agents* (pp. 227–252). Canberra: ANU.

Bernal-Agustín, J. L., Contreras, J., Martín-Flores, R., & Conejo, A. J. (2007). Realistic electricity market simulator for energy and economic studies. *Electric Power Systems Research*, 77, 46–54.

Bin, Z., Maosong, Y., & Xianya, X. (2004). The comparisons between pricing methods on pool-based electricity market using agent-based simulation. In *Proceedings of the IEEE international conference on electric utility deregulation, restructuring and power technologies (DRPT2004), hong kong*.

Binmore, K., & Swierzbinski, J. (2000). Treasury auctions: Uniform or discriminatory? *Review of Economic Design*, 5, 387–410.

Blackburn, J. M. (1936). *Acquisition of skill: An analysis of learning curves*. Industrial Health Research Board Report No. 73.

BMU. (2004). *Nationaler Allokationsplan für die Bundesrepublik Deutschland 2005-2007*. Bundesministerium für Umwelt, Naturschutz und Reaktorsicherheit, Berlin.

BMU. (2007). *Revidierter Nationaler Allokationsplan 2008–2012 für die Bundesrepublik Deutschland*. Bundesministerium für Umwelt, Naturschutz und Reaktorsicherheit, Berlin.

BMWI. (2007). *Energiedaten - nationale und internationale Entwicklung* (Tech. Rep.). Berlin: Federal Ministry of Economics and Technology, last updated Jun 13, 2007.

Bower, J., & Bunn, D. (2001). Experimental analysis of the efficiency of uniform-price versus discriminatory auctions in the England and Wales electricity market. *Journal of Economic Dynamics and Control*, 25, 561–592.

Bower, J., & Bunn, D. W. (2000). Model-based comparisons of pool and bilateral markets for electricity. *Energy Journal*, 21(3), 1–29.

Bower, J., Bunn, D. W., & Wattendrup, C. (2001). A model-based analysis of strategic consolidation in the german electricity industry. *Energy Policy*, 29(12), 987–1005.

Brenner, T. (2006). Agent learning representation: Advice on modelling economic learning. In L. Tesfatsion & K. L. Judd (Eds.), *Handbook of computational economics, volume 2: Agent-based computational economics* (pp. 895–947). Amsterdam: North-Holland.

Bundesnetzagentur. (2006). *Adjudication about the procedure of minute reserve procurement in Germany*. 29. August 2006, Bonn.

Bunn, D. W., & Day, C. J. (2002). *Computational modelling of price-formation in the electricity pool of England and Wales*. London Business School Decision Sciences Working Paper.

Bunn, D. W., & Oliveira, F. S. (2001). Agent-based Simulation: An application to the new electricity trading arrangements of England and Wales. *IEEE Transactions on Evolutionary Computation, Special issue: Agent Based Computational Economics*, 5(5), 493–503.

Bunn, D. W., & Oliveira, F. S. (2003). Evaluating individual market power in electricity markets via agent-based simulation. *Annals of Operations Research*, 121(1–4), 57–77.

Bunn, D. W., & Oliveira, F. S. (2007). Agent-based analysis of technological diversification and specialization in electricity markets. *European Journal of Operational Research*, 181(3), 1265–1278.

Bush, R., & Mosteller, F. (1955). *Stochastic models for learning*. New York: Wiley.

Camerer, C., & Ho, T.-H. (1999). Experienced-weighted attraction learning in normal form games. *Econometrica*, 67(4), 827–874.

References

Cau, T. D. H. (2003). *Analyising tacit collusion in oligopolistic electricity markets using a co-evolutionary approach.* Unpublished doctoral dissertation, Australian Graduate School of Management.

Cau, T. D. H., & Anderson, E. J. (2002). A co-evolutionary approach to modelling the behaviour of participants in competitive electricity markets. *IEEE Power Engineering Society Summer Meeting, 3,* 1534–1540.

Chao, H.-P., & Wilson, R. (2002). Multi-dimensional procurement auctions for power reserves: Robust incentive-compatible scoring and settlement rules. *Journal of Regulatory Economics, 22*(2), 161–183.

Cincotti, S., & Guerci, E. (2005). Agent-based simulation of power exchange with heterogeneous production companies. In *Computing in economics and finance 2005, society for computational economics.* (available at http://ideas.repec.org/p/sce/scecf5/334.html)

Cincotti, S., Guerci, E., Ivaldi, S., & Raberto, M. (2006). Discriminatory versus uniform electricity auctions in a duopolistic competition scenario with learning agents. In *IEEE congress on evolutionary computation,* Vancouver, Canada.

Cincotti, S., Guerci, E., & Raberto, M. (2005). Price dynamics and market power in an agent-based power exchange. In D. Abbott, J.-P. Bouchaud, X. Gabaix, & J.-L. McCauley (Eds.), *Noise and fluctuations in econophysics and finance; proceedings of the SPIE* (Vol. 5848, pp. 233–240).

Conlisk, J. (1996). Why bounded rationality? *Journal of Economic Literature, XXXIV,* 669–700.

Conzelmann, G., Boyd, G., Koritarov, V., & Veselka, T. (2005). Multi-agent power market simulation using EMCAS. In *IEEE power engineering society general meeting* (Vol. 3, pp. 2829–2834).

Cramton, P. (2003). Electricity market design: The good, the bad, and the ugly. In *Proceedings of the hawaii international conference on system sciences.*

Curzon Price, T. (1997). Using co-evolutionary programming to simulate strategic behaviour in markets. *Journal of Evolutionary Economics, 7,* 219–254.

Dales, J. H. (1968). *Pollution, property and prices: An essay in policy making and economics.* Toronto: University of Toronto Press.

Day, C. J., & Bunn, D. W. (2001). Divestiture of generation assets in the electricity pool of England and Wales: A computational approach to analyzing market power. *Journal of Regulatory Economics, 19*(2), 123–141.

dena. (2005). *Energiewirtschaftliche Planung für die Netzintegration von Windenergie in Deutschland an Land und Offshore bis zum Jahr 2020.* Expert's report by order of Deutsche Energie-Agentur GmbH (dena), Berlin.

Diekmann, J., & Schleich, J. (2006). Auktionierung von Emissionsrechten – Eine Chance für mehr Gerechtigkeit und Effizienz im Emissionshandel. *Zeitschrift für Energiewirtschaft ZfE, 30*(4), 259–266.

Dinther, C. van. (2007). *Adaptive bidding in single-sided auctions under uncertainty – an agent-based approach in market engineering.* Berlin: Springer.

Drogoul, A., Vanbergue, D., & Meurisse, T. (2003). Multi-agent based simulation: Where are the agents? In J. S. Sichman, F. Bousquet, & P. Davidsson (Eds.), *Multi-agent-based simulation, proceedings of the third international workshop mabs 2002, revised papers* (Vol. 2581, pp. 1–15). Springer.

Duffy, J. (2006). Agent-based models and human subject experiments. In L. Tesfatsion & K. L. Judd (Eds.), *Handbook of computational economics, volume 2: Agent-based computational economics* (pp. 949–1011). Amsterdam: North-Holland.

EEX. (2007). *Introduction to exchange trading at EEX on Xetra and Eurex.* http://www.eex.de. Leipzig.

EGL. (2006). *Price trend in Europe: Electricity markets booming.* Elektrizitäts-Gesellschaft Laufenburg AG, http://staticweb.egl.ch/eglgb/0506/en/preisentwicklungeu.html, accessed on Dec 17, 2007.

Ehlen, M. A., & Scholand, A. (2005). Modeling interdependencies between power and economic sectors using the N-ABLE agent-based model. In *Proceedings of the IEEE power engineering society general meeting* (pp. 2842–2846).

Ehlen, M. A., Scholand, A. J., & Stamber, K. L. (2007). The effects of residential real-time pricing contracts on transco loads, pricing, and profitability: Simulations using the N-ABLE™ agent-based model. *Energy Economics*, 29, 211–227.
EnWG. (2005). *Gesetz über die Elektrizitäts- und Gasversorgung*. BGBl I 2005, pp. 1970, 13. July 2005.
Epstein, J. M., & Axtell, R. L. (1996). *Growing artificial societies: Social science from the bottom up*. Cambridge: MIT.
Erev, I., & Roth, A. E. (1998). Predicting how people play games: Reinforcement learning in experimental games with unique, mixed-strategy equilibria. *American Economic Review*, 88(4), 848–881.
Ernst, D., Minoia, A., & Ilić, M. (2004a). Market dynamics driven by the decision-making of both power producers and transmission owners. In *IEEE power engineering society general meeting* (Vol. 1, pp. 255–260).
Ernst, D., Minoia, A., & Ilić, M. (2004b). Market dynamics driven by the decision-making of power producers. In *Proceedings of bulk power system dynamics and control* (Vol. IV). Cortina d'Ampezzo, Italy.
European Union. (1996). *Directive 96/92/EC of the European Parliament and of the Council of 19 December 1996 concerning common rules for the internal market in electricity*. Official Journal of the European Union, L 027.
European Union. (2003). *Directive 2003/87/EC of the European Parliament and of the Council of 13 October 2003 establishing a scheme for greenhouse gas emission allowance trading within the Community*. Official Journal of the European Union, L 275/32.
EWI/Prognos. (2007). *The trend of energy markets up to the year 2030* (Tech. Rep.). Berlin: by order of Federal Ministry of Economics and Technology.
Fagiolo, G., Birchenhall, C., & Windrum, P. (2007). Empirical validation in agent-based models: Introduction to the special issue. *Computational Economics*, 30(3), 189–194.
Franklin, S., & Graesser, A. (1997). Is it an agent or just a program? A taxonomy for autonomous agents. In J. Müller, M. Wooldridge, & N. Jennings (Eds.), *Intelligent agents iii: Agent theories, architectures, and languages, ECAI'96 workshop proceedings* (pp. 21–35). Springer.
Fudenberg, D., & Levine, D. (1998). *The theory of learning in games*. Cambridge: MIT.
Gilbert, N., & Bankes, S. (2002). Platforms and methods for agent-based modeling. In *Proceedings of the national academy of sciences of the united states of america* (Vol. 99, pp. 7197–7198).
Gilbert, N., & Troitzsch, K. G. (2005). *Simulation for the social scientist* (second ed.). Buckingham: Open University Press.
Gode, D. K., & Sunder, S. (1993). Allocative efficiency of markets with zero-intelligence traders: Market as a partial substitute for individual rationality. *The Journal of Political Economy*, 101(1), 119–137.
Graichen, P., & Requate, T. (2005). Der steinige Weg von der Theorie in die Praxis des Emissionshandels: Die EU-Richtlinie zum CO_2-Emissionshandel und ihre nationale Umsetzung. *Perspektiven der Wirtschaftspolitik*, 6(1), 41–56.
Greenwald, A. R., Kephart, J. O., & Tesauro, G. J. (1999). Strategic pricebot dynamics. In *EC '99: Proceedings of the 1st ACM conference on electronic commerce* (pp. 58–67). New York: ACM.
Gulyás, L. (2002). On the transition to agent-based modeling: Implementation strategies from variables to agents. *Social Science Computer Review*, 20(4), 389–399.
Guo, M., Liu, Y., & Malec, J. (2004). A new Q-learning algorithm based on the metropolis criterion. *IEEE Transactions on Systems, Man, and Cybernetics*, 34(5), 2140–2143.
Hämäläinen, R. P., Mäntysaari, J., Ruusunen, J., & Pineau, P.-O. (2000). Cooperative consumers in a deregulated electricity market – dynamic consumption strategies and price coordination. *Energy*, 25, 857–875.
Harp, S. A., Brignone, S., Wollenberg, B. F., & Samad, T. (2000). SEPIA – A simulator for electric power industry agents. *IEEE Control Systems Magazine*, 20(4), 53–69.
Hirschhausen, C. von, Weigt, H., & Zachmann, G. (2007). *Preisbildung und Marktmacht auf den Elektrizitätsmärkten in Deutschland – Grundlegende Mechanismen und empirische Evidenz*.

References

Expert's report by order of Verband der Industriellen Energie- und Kraftwirtschaft e.V. (VIK), Dresden.

Holland, J. H. (1975). *Adaptation in natural and artificial systems*. Reprint from 1992, Cambridge: MIT.

Holland, J. H., & Miller, J. H. (1991). Artificial adaptive agents in economic theory. *The American Economic Review, 81*(2), 365–370.

Jung, C., Krutilla, K., & Boyd, R. (1996). Incentives for advanced pollution abatement technology at the industry level: An evaluation of policy alternatives. *Journal of Environmental Economics and Management, 30*(1), 95–111.

Kaelbling, L. P., Littman, M. L., & Moore, A. W. (1996). Reinforcement learning: A survey. *Journal of Artificial Intelligence Research, 4*, 237–285.

Kahn, A. E., Cramton, P. C., Porter, R. H., & Tabors, R. D. (2001). Uniform pricing or pay-as-bid pricing: A dilemma for California and beyond. *The Electricity Journal, 14*(6), 70–79.

Kemfert, C., Schneider, F., & Wegmayr, J. (2007). Der Emissionshandel in Deutschland und Österreich - ein wirksames Instrument des Klimaschutzes? Discussion paper 28, Energieinstitut, University of Linz.

Klemperer, P. D. (2002). What really matters in auction design. *The Journal of Economic Perspectives, 16*(1), 169–189.

Klemperer, P. D., & Meyer, M. A. (1989). Supply function equilibria in oligopoly under uncertainty. *Econometrica, 57*(6), 1243–1277.

Koesrindartoto, D. (2002). *Discrete double auctions with artificial adaptive agents: A case study of an electricity market using a double auction simulator*. Department of Economics Working Papers Series, Working Paper 02005.

Koesrindartoto, D., Sun, J., & Tesfatsion, L. (2005). An agent-based computational laboratory for testing the economic reliability of wholesale power market desings. In *Proceedings of the IEEE power engineering society general meeting* (Vol. 3, pp. 2818–2823).

Koesrindartoto, D., & Tesfatsion, L. (2004). Testing the reliability of FERC's wholesale power market platform: An agent-based computational economics approach. In *Proceedings of the 24th USAEE/IAEE North American conference*. Washington, D.C., USA.

Krause, T., & Andersson, G. (2006). Evaluating congestion management schemes in liberalized electricity markets using an agent-based simulator. In *Proceedings of the power engineering society general meeting*. Montreal.

Krause, T., Andersson, G., Ernst, D., Beck, E. V., Cherkaoui, R., & Germond, A. (2005). A comparison of Nash equilibria analysis and agent-based modelling for power markets. In *Proceedings of the power systems computation conference (PSCC)*. Liege, Belgium.

Lane, D., Kroujiline, A., Petrov, V., & Sheblé, G. (2000). Electricity market power: marginal cost and relative capacity effects. In *Proceedings of the 2000 congress on evolutionary computation* (Vol. 2, pp. 1048–1055). La Jolla, USA.

Law, A. M. (2007). *Simulation modeling and analysis* (fourth ed.). New York: McGraw-Hill.

LeBaron, B. (2006). Agent-based computational finance. In L. Tesfatsion & K. L. Judd (Eds.), *Handbook of computational economics, volume 2: Agent-based computational economics* (pp. 1187–1233). Amsterdam: North-Holland.

Macal, C. M., & North, M. J. (2005). *Validation of an agent-based model of deregulated electric power markets* (Tech. Rep.). Los Alamos National Laboratory, presented at the North American Association for Computational and Social Organization (NAACSOS) Conference.

Machat, M., & Werner, K. (2007). Entwicklung der spezifischen Kohlendioxid-Emissionen des deutschen Strommix. *Climate Change (Umweltbundesamt), 01–07*.

Marks, R. E. (2006). Market design using agent-based models. In L. Tesfatsion & K. L. Judd (Eds.), *Handbook of computational economics, volume 2: Agent-based computational economics* (pp. 1339–1380). Amsterdam: North-Holland.

Marks, R. E. (2007). Validating simulation models: A general framework and four applied examples. *Computational Economics, 30*(3), 265–290.

McAfee, R. P., & McMillan, J. (1987). Auctions and bidding. *Journal of Economic Literature, 25*(2), 699–738.

Midgley, D. F., Marks, R. E., & Kunchamwar, D. (2007). The building and assurance of agent-based models: An example and challenge to the field. *Journal of Business Research, 60*, 884–893.

Miller, J. H. (1986). *A genetic model of adaptive economic behavior*. Working Paper, Department of Economics, University of Michigan.

Milliman, S. R., & Prince, R. (1989). Firm incentives to promote technological change in pollution control. *Journal of Environmental Economics and Management, 17*(3), 247–265.

Monopolies Commission. (2006). *More competition in the services sector as well*. The Sixteenth Biennial Report 2004/2005, abbreviated version, Bonn.

Moss, S. Edmonds, B. (2005). Sociology and simulation: Statistical and qualitative cross-validation. American Journal of Sociology, *110*(4), 1095–1131.

Mount, T. D. (2000). Strategic behavior in spot markets for electricity when load is stochastic. In *Proceedings of the 33rd hawaii international conference on system sciences*.

Müller, M., Sensfuß, F., & Wietschel, M. (2007). Simulation of current pricing-tendencies in the german electricity market for private consumption. *Energy Policy, 35*(8), 4283–4294.

Naghibi-Sistani, M., Akbarzadeh-Tootoonchi, M., Javidi-Dashte Bayaz, M.Rajabi-Mashhadi, H. (2006). Application of Q-learning with temperature variation for bidding strategies in market-based power systems. *Energy Conversion and Management, 47*, 1529–1538.

Nicolaisen, J., Petrov, V., & Tesfatsion, L. (2001). Market power and efficiency in a computational electricity market with discriminatory double-auction pricing. *IEEE Transactions on Evolutionary Computation, 5*(5), 504–523.

Nicolaisen, J., Smith, M., Petrov, V., & Tesfatsion, L. (2000). Concentration and capacity effects on electricity market power. In *Proceedings of the 2000 congress on evolutionary computation* (Vol. 2, pp. 1041–1047). La Jolla, USA.

North, M. J., Collier, N. T., & Vos, J. R. (2006). Experiences creating three implementations of the repast agent modeling toolkit. *ACM Transactions on Modeling and Computer Simulation, 16*(1), 1–25.

Oren, S. S. (2001). Design of ancillary service markets. In *Proceedings of the 34th Hawaii international conference on system sciences*.

Oren, S. S., & Sioshansi, R. (2004). Joint energy and reserves auction with opportunity cost payment for reserves. In *Bulk Power System Dynamics and Control* (Vol. VI).

Petrov, V., & Sheblé, G. (2001). Building electric power auctions with improved Roth-Erev reinforced learning. In *Proceedings of the North American power symposium*. Texas, USA.

Petrov, V., & Sheblé, G. B. (2000). Power auctions bid generation with adaptive agents using genetic programming. In *Proceedings of the 2000 North American power symposium*. Institute of Electrical and Electronic Engineers, Waterloo-Ontario, Canada.

Phan, D. (2004). From agent-based computational economics towards cognitive economics. In P. Bourgine & J. Nadal (Eds.), *Cognitive economics* (pp. 371–398). Berlin: Springer.

Praça, I., Ramos, C., Vale, Z., & Cordeiro, M. (2004). Intelligent agents for the simulation of competitive electricity markets. *International Journal of Modelling and Simulation, 2*, 73–79.

Pyka, A., & Fagiolo, G. (2005). *Agent-based modelling: a methodology for neo-schumpeterian economics* (Discussion Paper Series No. 272). Universitaet Augsburg, Institute for Economics. (available at http://ideas.repec.org/p/aug/augsbe/0272.html)

Railsback, S. F., Lytinen, S. L., & Jackson, S. K. (2006). Agent-based simulation platforms: Review and development recommendations. *Simulation, 82*(9), 609–623.

Rapoport, A., Seale, D. A., & Winter, E. (2000). An experimental study of coordination and learning in iterated two-market entry games. *Economic Theory, 16*, 661–687.

Rassenti, S. J., Smith, L. V., & Wilson, B. J. (2003). Controlling market power and price spikes in electricity networks: Demand-side bidding. *Proceedings of the National Academy of Sciences of the United States of America, 100*(5), 2998–3003.

Richiardi, M. (2004). *The promises and perils of agent-based computational economics*. EconWPA Working Paper, Report-no mgr040107.

Richiardi, M., Leombruni, R., Saam, N., & Sonnessa, M. (2006). A common protocol for agent-based social simulation. *Journal of Artificial Societies and Social Simulation, 9*(1).

Richter, C. W., & Sheblé, G. B. (1998). Genetic algorithm evolution of utility bidding strategies for the competitive marketplace. *IEEE Transactions on Power Systems, 13*(1), 256–261.
Roop, J. M., Fathelrahman, E., & Widergren, S.(2005). Price response can make the grid robust: an agent-based discussion. In *Proceedings of the IEEE power engineering society general meeting* (Vol. 3, pp. 2813–2817).
Roop, J. M., & Fathelrahman, E. (2003). Modeling electricity contract choice: An agent-based approach. In *Proceedings of the ACEEE summer study meeting.* Rye Brook, New York.
Roth, A. E. (2002). The economist as engineer: Game theory, experimentation, and computation as tools for design economics. *Econometrica, 70,* 1341–1378.
Rothkopf, M. H.(1999). Daily repetition: A neglected factor in the analysis of electricity auctions. *The Electricity Journal, 12*(3), 60–70.
Rupérez Micola, A., Banal Estañol, A., & Bunn, D. W. (2006). *Incentives and coordination in vertically related energy markets.* Discussion Paper SP II 2006 – 02, Wissenschaftszentrum Berlin.
Sargent, R. G. (2005). Verification and validation of simulation models. In *Proceedings of the winter simulation conference.*
Schummer, J., & Vohra, R. V.(2003). Auctions for procuring options. *Operations Research, 51*(1), 41–51.
Selten, R.(1991). Evolution, learning, and economic behavior. *Games and Economic Behavior, 3,* 3–24.
Simon, H. A.(1955). A behavioral model of rational choice. *Quarterly Journal of Economics, 69,* 99–118.
Stoft, S., Belden, T., Goldman, C., & Pickle, S. (1998). *Primer on electricity futures and other derivatives* (Tech. Rep.). University of California at Berkeley, Environmental Energy Technologies Division, No. LBNL-41098.
StromNEV.(2005. Verordnung über die Entgelte für den Zugang zu Elektrizitätsversorgungsnetzen (Stromnetzentgeltverordnung). BGBl I 2005, pp. 2225, 29. July 2005.
StromNZV.(2006). *Verordnung über den Zugang zu Elektrizitätsversorgungsnetzen (Stromnetzzugangsverordnung).* BGBl I 2006, pp. 2243, 29. July 2006.
Sun, J., & Tesfatsion, L. (2007). Dynamic testing of wholesale power market designs: An open-source agent-based framework. *Computational Economics, 30*(3), 291–327.
Sutton, R. S., & Barto, A. G.(1998). *Reinforcement learning: An introduction.* Cambridge: MIT.
Swider, J. D.(2006). *Handel an Regelenergie- und Spotmärkten: Methoden zur Entscheidungsunterstüutzung für Netz- und Kraftwerksbetreiber.* Wiesbaden: Deutscher Universitäts-Verlag.
Tesfatsion, L.(2002). Agent-based computational economics: Growing economies from the bottom up. *Artificial Life, 8,* 55–82.
Tesfatsion, L.(2006). Agent-based computational economics: A constructive approach to economic theory. In L. Tesfatsion & K. L. Judd (Eds.), *Handbook of computational economics, volume 2: Agent-based computational economics* (pp. 831–880). Amsterdam: North-Holland.
Thorndike, E. L. (1898). Animal intelligence: An experimental study of the associative processes in animals. *Psychological Monographs, 2*(8).
Tobias, R., & Hofmann, C.(2004). Evaluation of free Java-libraries for social-scientific agent based simulation. *Journal of Artificial Societies and Social Simulation, 7*(1).
UCTE.(2006). *Operation handbook.* Brussels, Belgium.
VDEW.(1999). *Repräsentative VDEW-Lastprofile, M-28/99.*
VDN. (2004). *Leistungsbilanz der allgemeinen Stromversorgung in Deutschland: Vorschau 2005 - 2015* (Tech. Rep.). Berlin: Verband der Netzbetreiber – VDN e.V., part of VDEW.
VDN. (2007. TransmissionCode 2007 – Netz- und Systemregeln der deutschen Übertragungsnetzbetreiber, Version 1.1. Version 1.1, Verband der Netzbetreiber – VDN e.V., part of VDEW. Berlin.
Veit, D. J., Weidlich, A., Yao, J., & Oren, S. S.(2006). Simulating the dynamics in two-settlement electricity markets via an agent-based approach. *International Journal of Management Science and Engineering Management, 1*(2), 83–97.

Ventosa, M., Báillo, A., Ramos, A., & Rivier, M. (2005). Electricity market modeling trends. *Energy Policiy, 33*, 897–913.
Visudhiphan, P. (2003). *An agent-based approach to modeling electricity spot markets*. Unpublished doctoral dissertation, Massachusetts Institute of Technology.
Visudhiphan, P., & Ilić, M. D. (1999). Dynamic games-based modeling of electricity markets. In *Power engineering society 1999 winter meeting, IEEE* (Vol. 1, pp. 274–281).
Visudhiphan, P., & Ilić, M. D. (2001). An agent-based approach to modeling electricity spot markets. In *Proceedings of IFAC modeling and control of economic systems* (pp. 407–412). Klagenfurt, Austria.
Visudhiphan, P., & Ilić, M. D. (2002). On the necessity of an agent-based approach to assessing market power in the electricity markets. In *Proceedings of the tenth international symposium on dynamic games and applications* (Vol. 2). Sankt Petersburg, Russia.
Watkins, C. J. C. H. (1989). *Learning from delayed rewards*. Unpublished doctoral dissertation, Cambridge.
Weidlich, A., & Veit, D. (2006). Bidding in interrelated day-ahead electricity markets: Insights from an agent-based simulation model. In *Proceedings of the 29th IAEE international conference*. Potsdam.
Weidlich, A., & Veit, D. (2008a). Agent-based simulations for electricity market regulation advice: Procedures and an example. *Journal of Economics and Statistics, 228*(2+3), 149–172.
Weidlich, A., & Veit, D. (2008b). Analyzing interrelated markets in the electricity sector – The case of wholesale power trading in Germany. In *IEEE Power Engineering Society General Meeting*, Pittsburg, USA.
Weidlich, A., & Veit, D. (2008c). A critical survey of agent-based wholesale electricity market models. *Energy Economics, 30*(4), 1728–1759.
Weidlich, A., & Veit, D.(2008d). PowerACE: Ein agentenbasiertes Tool zur Simulation von Strom- und Emissionsmärkten. In *Proceedings of the multikonferenz wirtschaftsinformatik, track: "it in der energiewirtschaft"*. Garching.
Weinhardt, C., Holtmann, C., & Neumann, D.(2003). Market engineering. *Wirtschaftsinformatik, 45*(6), 635–640.
Werker, C., & Brenner, T. (2004). *An advanced methodology for heterodox simulation models based on critical realism*. Working Paper 0401, Papers on Economics and Evolution, Max Planck Institute of Economics, Evolutionary Economics Group, Jena, Germany.
Wilson, R. (1985). Incentive efficiency of double auctions. *Econometrica, 53*(5), 1101–1115.
Wilson, R. (2002). Architecture of power markets. *Econometrica, 70*, 1299–1340.
Windrum, P., Fagiolo, G., & Moneta, A. (2007). Empirical validation of agent-based models: Alternatives and prospects. *Journal of Artificial Societies and Social Simulation, 10*(2, 8).
Xiong, G., Okuma, S., & Fujita, H. (2004). Multi-agent based experiments on uniform price and pay-as-bid electricity auction markets. In *Proceedings of the IEEE international conference on electric utility deregulation, restructuring and power technologies (DRPT2004)*. Hong Kong.
Yu, J., Zhou, J.-Z., Yang, J., Wu, W., Fu, B., & Liao, R.-T. (2004). Agent-based retail electricity market: Modeling and analysis. In *Proceedings of the third international conference on machine learning and cybernetics*. Shanghai.